Math Squared

Graph Paper Activities
for Fun and Fundamentals

Math Squared

Graph Paper Activities
for Fun and Fundamentals

DAVID STERN

Teachers College, Columbia University
New York & London 1980

Copyright © 1980 by Teachers College, Columbia University
Published by Teachers College Press, 1234 Amsterdam Avenue,
New York, N.Y. 10027

Library of Congress Cataloging in Publication Data

Stern, David Peter, 1931–
    Math squared.

    Includes index.
    SUMMARY: Brief discussions of mathematical ideas,
with problems and puzzles to be worked out with graph
paper. Solutions are included.
    1. Mathematical recreations.   2. Graphic methods.
[1. Mathematical recreations.   2. Graphic methods]
I. Title.   II. Title: Graph paper activities for
fun and fundamentals.
QA95.S72   793.7′4   80-15932
ISBN 0-8077-2585-4 (pbk.)

Manufactured in the United States of America.
10 9 8 7 6 5 4 1   80 81 82 83 84 85

Contents

52	61	4	13	20	29	36	45
14	3	62	51	46	35	30	19
53	60	5	12	21	28	37	44
11	6	59	54	43	38	27	22
55	58	7	10	23	26	39	42
9	8	57	56	41	40	25	24
50	63	2	15	18	31	34	47
16	1	64	49	48	33	32	17

A "Magic Square" by Benjamin Franklin

1

Introduction

In many countries outside the U.S.A., students are required to do all their math work on "square-ruled paper": paper provided with ruled squares. It is sometimes called "quadrille paper" (a quadrille long ago was a square formation in which knights paraded—and, later, a dance performed in a square formation).

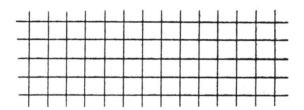

In countries where such paper is used for calculations, some teachers seem to believe it helps students by keeping additions and subtractions lined up, without the tens getting into the column meant for the hundreds, and all that. Actually, of course, one soon finds that there is much more to math than keeping one's columns straight—which may be the reason why some students seem to manage their math problems just as easily (or with just as much difficulty) on plain paper.

Yet those little squares can do much *more* than keep columns straight. They can help make math more interesting and more understandable, in many ways. I myself attended school abroad, and I recall being intrigued and helped by a great number of puzzles, games, and mathematical "side trips," all of which used square-ruled paper—or, as it is more often called in the U.S.A., "graph paper." This little book is meant to introduce you to some of them, and I wish you as much fun as I had.

In the chapters that follow you will find a wide variety of subjects connected in one way or another (sometimes quite loosely, I confess) with graph paper. All the reader will need beforehand is some basic knowledge of arithmetic, up to and including operations with fractions, and a reasonably good reading ability.

This book is intended for two rather different groups of readers. On the one hand, it is meant for advanced students of 7th-to-9th-grade mathematics—students who find the standard curriculum too limited and would like some new and more challenging material. Sometimes such students are handed material intended for later school years, but this merely postpones their problem. Here a different option is offered— the option of exploring a broader range of subjects, including some areas rarely discussed in the classroom.

On the other hand, adults seem to enjoy the book as well. Some of my associates have read it in draft form, and even those among them who had a background in the technical professions always discovered entertaining twists and byways that were not familiar to them. Some adult readers were attracted by the collection of unusual facts and puzzles, and many who had moved away from math after leaving school found here new interest and a fresh viewpoint.

What the book contains is a quick tour showing the mathematical beginner or amateur what math really *is*—not the shuffling of numbers or the memorizing of formulas, but the development of *ideas*. You will read here about graphs and formulas, street plans and bridges, furlongs and barleycorns—and also about Pythagoras and Gauss, and Mark Twain and Ben Franklin. The aim is to show the many and various forms that a mathematical idea can have and the many different ways in which math is related to everyday life and culture. And what you read (and *do*) may even increase your appetite—and your ability—to learn more on your own.

A few words about the problems and puzzles scattered throughout the book (signaled by question marks in the margin). The answers are given in other places (indicated by exclamation marks), but try not to

peek: work out the solutions by yourself (if you can at all), because this way you get the satisfaction of discovering things on your own. On the other hand, you may skip anything that seems too difficult. This is not a mathematical textbook, and you do not have to master every subject in order to understand what follows. To get the most out of the book, make sure you have plenty of graph paper handy, so you can try things out by yourself while reading.

This is meant to be a fun book—the kind that might be appreciated on a long trip or a rainy weekend. In writing it, I have tried to imagine that I was telling it to my own children, who seem to have a lot of fun with math. Technical wording has been avoided as much as possible, and the mathematics, too, is often simplified. This is all right for a first look at the subject: if your interest continues, you will probably come back to these matters, some day, and study them in more detail.

And now, to begin—a puzzle.

2

The Prisoner's Escape

Let us call a "unit" the width of each small square on the ruled paper (more will be said about that "unit" later on). We now draw a big square, eight units wide and eight units high, with a small opening at the top left corner.

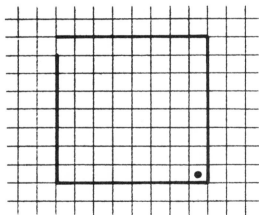

The drawing is the map of a jail: it contains 64 rooms, or "cells"—every square is a room—and there exist doors in the walls between any two neighboring rooms (that is, you can move from one to another going up, down, left, or right, but *not* diagonally through a corner). In addition there is just the *one* door leading *out* of the jail: the opening at the top left corner.

A prisoner sitting in the bottom right corner cell (marked by a dot) is told that he may leave the jail and go free *if* on his way out he visits every other cell once and no more than once (his own cell he may enter as many times as he wishes). He may open any door to accomplish his pur-

4

pose, but at the end of his trip he must arrive at the prison's exit, where he would be allowed to continue out to his freedom.

How does he move?

Before you continue, get hold of a sheet of square-ruled paper (or draw your own rulings) and try to trace the correct route.

• • • • •

If you have solved the problem, congratulations! Either you are unusually sharp—or you have seen it before.

If you haven't succeeded, you probably found that one cell was always left over, as shown below:

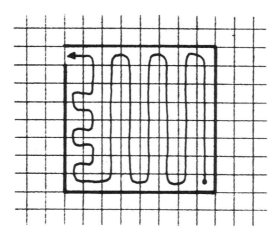

To see why this happens, let us shade the cells in checkerboard fashion and note the color of each cell visited by the prisoner:

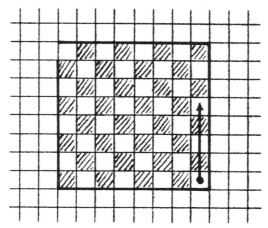

The first cell (his own) is white.
The second one is black.
The 3rd one is white.
The 4th one is black.

And so on and so forth: the color always changes, for from a white cell the prisoner can only enter black cells, and from a black cell only white ones.

Therefore, no matter how he moves, all *even-numbered* cells are black, all *odd-numbered* ones are white.

If the prisoner is to end his trip next to the exit gate, the last cell he visits is the one in the top left corner, which is white. *But*—if he has visited every cell once and this is the last one, it will be his 64th visit; and since 64 is an even number, the cell ought to be *black*.

So it appears as if the problem cannot be solved. The prisoner can easily reach the white exit cell on his 63rd visit (63 is odd) but if he does so, somewhere behind him will be left one unvisited black cell.

Yet there *is* a way, though it requires a certain trick. The solution is given on page 8, but before you look it up, read the *problem* again—carefully—and see if you can discover the trick yourself.

3

Rectangles

We now turn to something simpler—to some of the shapes that can be drawn with the help of square ruling on paper.

Simplest among such shapes are *rectangles*, and it is easier to draw one than to describe it:

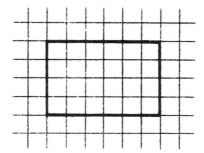

Rectangles come in all sizes, big and small. One way of measuring the size of a rectangle is by counting the number of squares it contains: we call this the *area* of the rectangle, and there is a simple rule for finding it quickly without counting. For example, the rectangle drawn here is 6 units wide and 4 units high: the rule says that we get the area if we *multiply* the width by the height; so the area must be

$$6 \times 4 = 24 \text{ squares.}$$

By the way, whenever areas are measured, one should always state the size of the "unit" in which distances are measured. Here the "unit" is the distance between two neighboring lines of the ruling (its relation to other units of length will be described in chapter 14). If, on the other hand, the unit of distance were chosen to be one inch, all areas would be given in "square inches."

It is easy to see why the size of the length unit is important. Suppose we had a square ruling with a unit five times smaller than the one used so far. Then a rectangle of the *same size* as the one drawn earlier would appear as shown below:

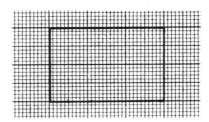

It is quite clear that the number of squares in this rectangle is much larger than 24, so the number giving the area in terms of the new units is now not 24 but a much larger number (600, in fact). In the rest of this book, the same "unit" will be used everywhere and for this reason we will not worry any more about it.

In mathematical language, the result of multiplying two numbers together is called their *product*. So the rule for finding the area of rectangles can be simply stated this way:

> The area of a rectangle is the product of its width and its height.

Solution

The prisoner's escape:

The prisoner may visit his own cell more than once. Until now this fact has not been used: let's see whether it provides any help.

Suppose that the prisoner's first visit is by the "top door" to the cell just above his own (in the drawing) and suppose that he *does* wish to revisit his own cell. He must do so *at once*: if he does not, then he *either* has to enter once more the cell above his own, and this is forbidden, since he can visit it only once; *or* he can enter his own cell by the other door, on its left side, and then he cannot leave again, since both neighboring cells have already been visited and cannot be entered a second time.

If, however, the prisoner visits the cell above his own, returns at once to his own cell and then leaves again by the remaining door on the left, his escape becomes quite easy. The reason is simple: the prisoner now makes a total of 65 visits (including two to his own cell) and the last cell can be white as required, since 65 is an odd number.

Try it and see!

Can any number be a product? For instance, is 79 the product of any two whole numbers? Or—in other words—can one draw a rectangle with an area of 79 without cutting through any squares?

The answer depends on the rules we follow. It is certainly correct to write

$$79 \times 1 = 79,$$

and therefore a rectangle of area 79 *can* be drawn, provided it is a long strip one unit wide and 79 units long. However, such products involving multiplication by 1 can be written for any number; so *let us not count them*.

It then turns out that 79 is the product of *no* two whole numbers. Numbers with this property are called *prime numbers* or, for short, *primes*. The number 79 is a prime, but its nearest neighbors are not:

$$78 = 6 \times 13.$$
$$80 = 8 \times 10.$$

Prime numbers are scattered among whole numbers like raisins in a pudding: they become somewhat more rare as one goes to bigger numbers, but they never end. The numbers 2, 3, 5, 7, 11, and 13 are prime; and some more prime numbers, following these, are given in a list on page 11. Before looking at it, you are invited to make up your own list and then compare!

4

Squares

A square is a special kind of rectangle—its width is the same as its height. Only one number is needed to tell everything about a square, and this will be called the *side* of the square—it can be either height or width, since these are equal.

Let us examine the squares whose side equals a whole number of units—and, in particular, find their areas:

A square 1 unit wide has an area $\quad 1 \times 1 = 1.$

$\qquad \ldots \qquad$ 2 units wide $\quad \ldots \qquad 2 \times 2 = 4.$

$\qquad \ldots \qquad$ 3 units wide $\quad \ldots \qquad 3 \times 3 = 9.$

$\qquad \ldots \qquad$ 4 units wide $\quad \ldots \qquad 4 \times 4 = 16.$

$\qquad \ldots \qquad$ 5 units wide $\quad \ldots \qquad 5 \times 5 = 25.$

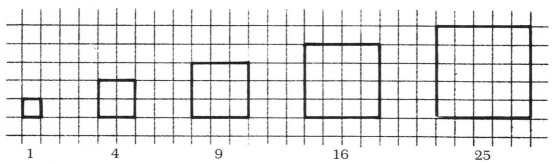

1   4   9   16   25

The next numbers on the list of areas would be

36, 49, 64, 81, 100, 121, 144, 169, 196, 225, 256, 289, 324;

and, of course, the list can be continued without limit.

10

This rather interesting group of numbers has a name: they are called *square numbers* or simply *squares*. Squares are always formed by multiplying a number by itself, and we speak of "the square of a number" meaning the product of a number with itself. For instance, saying "25 is the square of 5" means that

$$5 \times 5 = 25.$$

There is a short way to indicate that a number is multiplied by itself one or more times, or that a number is a product of two or more equal numbers. For example, we often write

$$5^2 \text{ instead of } 5 \times 5.$$

In other words 5^2 is the product of 2 fives. It is sometimes called "five to the second power"; but usually we call it "the square of five" or "five squared." One also writes

$$5^3 \text{ instead of } 5 \times 5 \times 5,$$

which is called "five to the third power" or simply "five to the third," or sometimes "five cubed." In the same manner,

$$2^4 \text{ means } 2 \times 2 \times 2 \times 2$$

and is called "two to the fourth power" or "two to the fourth." Some other powers written in this fashion include these:

$2^2 = 4.$	$3^2 = 9.$
$2^3 = 8.$	$3^3 = 27.$
$2^4 = 16.$	$3^4 = 81.$
$2^5 = 32.$	$3^5 = 243.$

Solution

The primes following 13 and smaller than 200:

17	19	23	29	31	37	41	43	47	53
59	61	67	71	73	79	83	89	97	101
103	107	109	113	127	131	137	139	149	151
157	163	167	173	179	181	191	193	197	199

The largest prime number known at the time I'm writing equals $2^{44497} - 1$. Written in the usual form, this number would be 13,395 digits long. Of course, the numbers that form the foundation of our everyday system for naming and writing down numbers—ten, hundred, thousand, ten thousand ("myriad" in the Bible), and so on through the million and the billion—are all powers of ten. However, we'd better stop here, since this is already pretty far from the subject of squares!

5

Formulas

Notice that the squares of numbers grow faster than the numbers themselves:

Number:	1	2	3	4	5	6	7	8	9	10
Its square:	1	4	9	16	25	36	49	64	81	100

While the numbers in the top row grow at a steady rate, getting larger by 1 at each step, their squares (in the lower row) increase at a rate that speeds up all the time: the first two squares in our list differ only by 3, but the last two differ by 19! Another sign showing that the squares of numbers grow faster than the numbers themselves is the fact that when a number doubles, its square increases not twice but *four* times: 6 is twice as large as 3, but $6^2 = 36$ is four times larger than $3^2 = 9$.

Many things are known to grow in this fashion. For instance, if a car accelerates to *twice* its speed, it becomes *four* times harder to stop it. Let's take a closer look at what exactly this means.

Suppose you are driving a car on a level dry road at 40 miles per hour—that is, a speed which brings you 40 miles farther down the road for each hour of driving—and suddenly you have to stop. You step on the brake—but of course the car does not stop immediately: some time is needed before it slows down and comes to a complete stop. How far will the car continue moving after you apply the brakes?

The exact answer of course depends on many things—on the car, its tires, the condition of the road, and so on. However, a fairly accurate answer can be found by the following rule. Take the number of tens of miles in your speed (here it is 4), form its square, and multiply everything by 5: the result is the "braking distance" in feet.

In the particular case of a car moving at 40 miles per hour, this means

$$4^2 \times 5 = 16 \times 5 = 80 \text{ feet.}$$

If instead you were driving at 80 miles per hour, the distance would be

$$8^2 \times 5 = 64 \times 5 = 320 \text{ feet.}$$

This is quite far: although the car is only moving twice as fast as before, it now goes *four* times farther before stopping. Obviously, driving twice as fast is more than twice as dangerous!

Mathematicians have a special way of writing down rules of this sort, by means of *formulas*. In a formula, any number which is not known beforehand is marked by a *letter*. For instance, the rule for the area of a rectangle can be written as a formula thus:

$$A = W \times H,$$

where W is the number giving the width,
   H is the number giving the height, and
   A is the number giving the area.

If you are told that in a particular case W equals 6 and H equals 4, you can replace the letters in the formula by these numbers and the formula then gives the correct number for A, namely

$$A = 6 \times 4 = 24.$$

The rule for the "braking distance" of a car can also be written as a formula, namely

$$D = N^2 \times 5,$$

where N stands for the number of tens of miles-per-hour on the car's speedometer and D is the braking distance in feet.

To *use* the formula, suppose the car travels at 50 miles per hour. Then $N = 5$, and by putting the number 5 in place of N in the formula we get the distance D as

$$D = 5^2 \times 5 = 125 \text{ feet.}$$

As the speed gets higher, the braking distance grows at an increasing rate, as can be seen from the table below:

Speed in Miles per Hour	D in feet
10	5
20	20
30	45
40	80
50	125
60	180
70	245
80	320

As mentioned before, this formula is not completely accurate: the exact braking distance depends on the roughness of the road, the condition of the car's tires and brakes, and other things. Also, it only applies to level dry roads: on a wet (or icy) road, or on one that slopes downward, stopping a car is much harder. On the other hand, the formula will also work if N is not a whole number—for 25 miles per hour, for instance, $N = 2\frac{1}{2}$ and $N^2 = 6\frac{1}{4}$.

6

An "Almost-Formula" for Prime Numbers

We turn to one more formula using squares.

No one has yet found a simple formula that derives prime numbers. Mathematicians have long searched for one, but the only general method for finding primes is still the careful examination of numbers to see whether they can be divided by smaller numbers—if they cannot, they are prime.

Perhaps the nearest thing to such a formula was discovered by the Swiss-born mathematician Leonhard Euler (pronounced "oiler") about 200 years ago. It is

$$P = N^2 + N + 41.$$

If N is replaced by a small whole number, P is always prime. However, the formula clearly does not hold without a limit: if $N = 41$, the number P is completely "made up" by adding and multiplying the number 41, and one would therefore suspect that it can be divided by 41 (indeed it can). In fact, even for $n = 40$ the result can be divided by 41; but if N is a whole number between (and including) 0 and 39, P is a prime number. The formula thus gives 40 prime numbers one after the other—a record for formulas as simple as this one.

By the way, the choice of the letter N to represent the unknown number in the above formula is not an accident. For some reason it has become customary in formulas to denote whole numbers by the letter N (capital or small): if several whole numbers are involved, the letters preceding N in the alphabet are also used, all the way down to I. In a widely used system for handling formulas by computers—known as the "computer language" FORTRAN, short for FORmula TRANslation—whenever a quantity is denoted by a letter from I to N (or by a group of letters *beginning* with a letter between I and N) the computer automatically assumes (unless it is instructed otherwise) that the quantity is a whole number; in other cases, it will provide a suitable decimal point. Unknown numbers which may or may not be whole are usually denoted in formulas by the letter X—and if several such numbers are involved, Y and Z are also used.

7

Irrational Numbers

It is possible to draw squares with areas that are not included in the list of square numbers, but the length of their side will not in general be a whole number of units.

For instance, the square

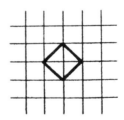

has an area of 2. (The reason for this is explained on page 20. Can you guess it?). Its *area* is a whole number of squares—two—but the length of its side is somewhere between one and two units. In fact, the side turns out to be just a bit less than 1½ units long.

The exact number giving the length of the side is called "the square root of two," which is another way of saying "the number the square of which equals two." There is a special symbol for writing it, namely

$$\sqrt{2}.$$

Obviously, other numbers will also have square roots. If a number belongs to the list of square numbers, its square root will be a whole number; the square root of 9, for instance, is 3:

$$\sqrt{9} = 3.$$

However, $\sqrt{3}$, $\sqrt{5}$, $\sqrt{6}$, $\sqrt{7}$ and most other square roots—including

$\sqrt{2}$—are not so simply expressed. They belong to an interesting (and very large) group of numbers called *irrational numbers*, which literally means "numbers that do not make sense." Their story is as follows.

Some 2500 years ago there lived a Greek mathematician named Pythagoras. He seems to be one of the first mathematicians whose names are known to us, and he lived at a time when very little was known about numbers and science.

Numbers, especially whole numbers, fascinated Pythagoras. He discovered interesting rules concerning them, as well as a famous formula about triangles which still bears his name. He felt that there was something particularly beautiful about whole numbers and that they held the key to understanding nature.

Now obviously not every number is whole—there also exist fractions and numbers with a whole and a fractional part, such as 1½. Still, Pythagoras believed that if a number is not whole, it can always be given by *two* whole numbers, one divided by the other. For instance, 1½ is 3 divided by 2, or $\frac{3}{2}$ for short (also written 3/2). This is nowadays called the "ratio" between 3 and 2—literally, "the sensible way" of combining 3 and 2 into a single number. (The word *ratio* comes from Latin, the language of the ancient Romans, which has contributed many other words to English. The Pythagoreans spoke Greek and used a different name, based on the Greek word *logos* from which the English "logic" is derived. However, their word meant the same as "ratio"—it was to them, you might say, "the logical way" of combining two numbers.)

Any whole number and any number containing a fractional part ("mixed number") can be written as a ratio. Furthermore, if you happen to be familiar with decimal fractions, you will realize that whenever such a fraction (or indeed any number which contains the decimal point) comes to an end on the right hand side of the decimal point, it too can be represented as a ratio. All such numbers are called *rational numbers*: for example, 3/2, 22/7, 355/113, 2.54 = 254/100, and 137 = 137/1 are all rational.

Pythagoras had students and friends to whom he taught his ideas and discoveries. Together they formed the "Pythagorean Brotherhood"—a society devoted to the study of numbers and of nature. For a while the society grew, in spite of some very strange beliefs held by its members. Then one day a member of the society made an unexpected discovery: contrary to what he had been taught, not *all* numbers could be expressed as the ratio of two whole numbers. In particular, the number $\sqrt{2}$ could not: one can find fractions that come as close to it as we please, but no fraction exists which gives $\sqrt{2}$ *exactly*.

The Pythagoreans called such numbers (in Greek) *alogos* which in English translates into "illogical" (numbers), and we now use a similar name of Latin (Roman) origin—"irrational numbers," meaning either numbers which cannot be expressed as a ratio or, if you wish, numbers which do not make sense. At one time they were called "surds" in English, not on account of their "absurd" nature (though that, too, could be claimed) but from the Latin *surdus*, meaning deaf. What happened was a slight error in translation: *alogos* also means "without a word" and this caused an Arab mathematician, about 1000 years ago, to translate it as "deaf"; later, upon further translation from Arabic into Latin, the language of scholars in the middle ages, this emerged as *surdus*!

It is told that members of the brotherhood were deeply upset by the discovery (one legend claims that they killed the discoverer in an attempt to keep the matter secret), and it led them to question the rest of their beliefs: the society broke up soon afterwards. Nowadays we know not only that irrational numbers do exist but also that in a way there are actually many *more* of such numbers than there are of the "ordinary" (or "rational") numbers of the kind that can be written as fractions. To show this, however, would be too complicated for this book.

The proof showing *why* the side of a square with area 2 is irrational is somewhat harder than the other discussions in this book and is therefore given in a separate section at the end.

Solution

Why the square has an area of 2:

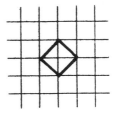

The lines of the ruling divide the square into 4 triangles, each equal to half of a ruled square. By putting the triangles together in a different way, two such squares can be constructed, and this therefore is also the area of the larger square.

8

Dividing Up Squares

Squares can be cut into parts in many different ways. For instance, one can divide a square into 4 equal parts like this:

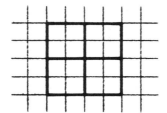

The big square has an area of $4^2 = 16$, so each smaller square must have an area one fourth of this, that is 4. In fact, no matter *how* one divides the square into 4 equal parts, each part must always have an area of 4.

See if you can solve these problems (the solutions are given later):

(1) Cut the square into 4 pieces each of which has this shape:

(If this seems too hard, a hint is given at the end of the list of problems.)

(2) Cut the square into 4 pieces each of which has this shape:

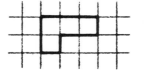

Can this be done in such a way that no two pieces, when fitted into the square, together form a rectangle?

21

(3) Show that it is *never* possible to cut the square into 4 pieces each of which is shaped like this:

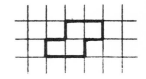

(4) Take away one corner of the big square, leaving the shape drawn below:

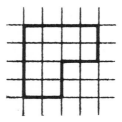

Can you divide this shape into 4 equal pieces, each of which has the same appearance as the big shape?

Hint for solving the first problem:
Each of the pieces has a long side of length 3 and several short sides of length 1. Imagine now that the big square is a box into which the 4 pieces must be fitted. The bottom of the box

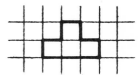

is 4 units long. Suppose first that *no* piece touches it with its long side. There is then only one way left for covering the bottom: by having every one of the 4 pieces touch it with one of its short sides. But such a solution could never work, since no piece would then be long enough to reach the top!

This leaves only one possibility: one piece *must* touch the bottom with its long side—

like this: 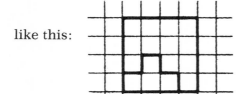 or like this: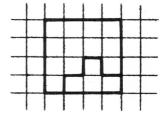

Fitting the remaining pieces is now quite easy. Two solutions are possible—depending on which of the above two positions you have started from—and you can get each of them from the other one by "flipping it over" (try looking at it against the light from the reverse side of the paper!).

9

Triangles

A triangle is a shape formed by three straight lines meeting at three corners, as shown here:

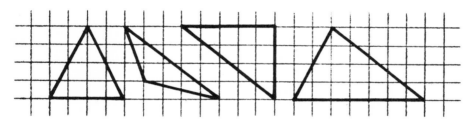

A corner between two straight lines is called an *angle*, and "triangle" therefore simply means a shape with three angles.

In particular, each of the four corners of a square forms what is called a *right angle*, and rectangles are so named because all their angles are of this kind. A triangle may have a right angle, but never more than one. Are there any right angles in the triangles drawn above?

Two lines that meet at a right angle are said to be *perpendicular* to each other. These two lines are perpendicular

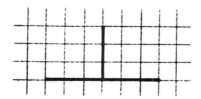

and so are any two edges of this page which meet at a corner.

Solutions

Dividing Up Squares:

Problem 1:

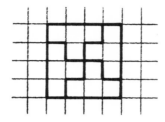

Problem 2:

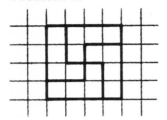

If pieces were allowed to form rectangles, this would be another solution:

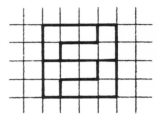

Problem 3:

Choose any corner square—for instance, the one in the bottom left corner. *Some* piece must cover it, and this can only be done in one of the two ways shown below:

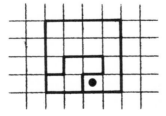

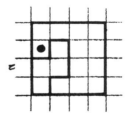

In either case, no second piece of the same shape can cover the square marked with a dot, so a solution is impossible.

Problem 4:

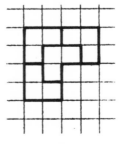

Let us draw a triangle so that one of its sides forms a level bottom. We now draw a line perpendicular to the bottom, beginning at the opposite corner—like the broken line in the drawing shown here:

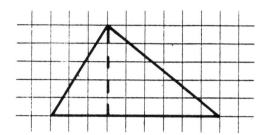

This is called *an altitude* (or height) of the triangle. Note that we say *an* altitude, not *the* altitude: by turning the same triangle around, each of the other two sides may be made the bottom, and a different altitude can then be drawn towards it from the opposing corner. A triangle therefore has a total of three altitudes. If a triangle leans over so that its top sticks out to one side, it is still possible to draw an altitude from the top corner; however, it will not meet the bottom of the triangle—only the continuation of the bottom, as shown in the following drawing.

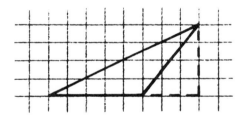

To sum up, you now know the meaning of these terms:

> Triangles
> Angles
> Right angles
> Lines perpendicular to each other
> The altitudes of a triangle

You are now ready to find out something less simple about the triangle: *what is its area?*

To find the answer, let the triangle again have one of its sides as a flat bottom—or, to use the proper mathematical name, as its *base*. We draw the altitude to this side and enclose the triangle in a rectangle—like this:

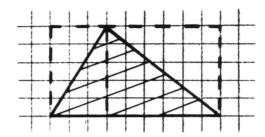

The triangle has *half* the area of the rectangle. To see why, it is best to pull the rectangle apart along the altitude of the triangle:

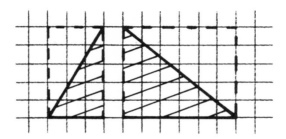

In each of the pieces, exactly *one half* belongs to the triangle, so the triangle must equal half the rectangle in area.

The *height* of the rectangle is the same as the altitude of the triangle, the *length* of the rectangle is the same as the length of the triangle's base: its *area* therefore equals the *product* of these two numbers. The area of the triangle is half of that, so:

Area of triangle = ½ × length of base × altitude.

It is interesting to note that one gets the same result no matter *which* of the three sides of the triangle is used as base. Of course, the altitude is usually different for each choice of base.

Our result is also correct for a triangle which leans over so that its altitude only meets the continuation of its base. Let the length of the altitude be denoted by the letter A and the length of the base by the letter B. Then, as the drawing shows, the unknown area X of the triangle can be obtained by subtracting the area of a triangle with base C from the area of a larger triangle with base $(B + C)$:

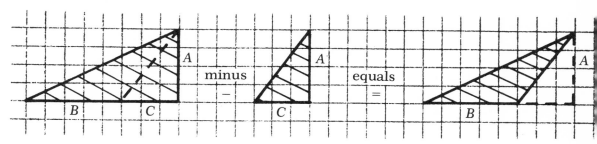

All the triangles here have the same altitude A, and therefore

$$X = \frac{1}{2} \times A \times (B + C) - \frac{1}{2} \times A \times C$$
$$= \frac{1}{2} \times A \times B + \frac{1}{2} \times A \times C - \frac{1}{2} \times A \times C$$
$$= \frac{1}{2} \times A \times B.$$

In the last equality we have dropped

$$\frac{1}{2} \times A \times C - \frac{1}{2} \times A \times C$$

since a number minus itself *always* amounts to zero. What remains is the same result which was derived earlier for "ordinary" triangles.

Using the above formula for the area of a triangle, one can find the area of any flat shape made up of straight lines, because such a shape can always be cut up into a number of triangles. For instance, an 8-sided shape similar to a STOP sign may be divided into six triangles in one of the ways shown here:

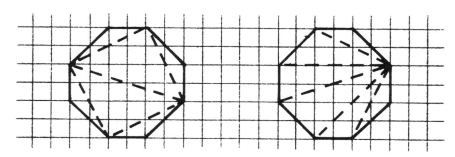

There exists a third way, one which can be drawn without lifting the pen from the paper and without touching any point twice. Can you guess it? The answer is on page 29.

10

Triangular Numbers

Small children often stack play blocks in triangle-shaped piles like this:

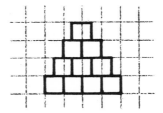

Each layer here has one block *less* than the one below it. Food stores also sometimes stack boxes or cans in such piles. When drawing a pile like this on square-ruled paper, one must cut through some of the squares; however, it is possible to draw triangle-shaped piles without cutting through squares if we shift the blocks so that one side of the pile is straight, *or* if we draw brick-shaped blocks:

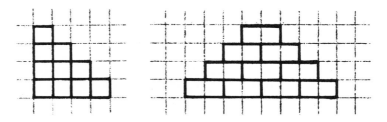

The number of "blocks" in any of the preceding drawings is

$$1 + 2 + 3 + 4 = 10.$$

It is the fourth in the series of "triangular numbers" ("triangular" means "related to triangles") which give the number of blocks in triangle-shaped piles. The first five such numbers are:

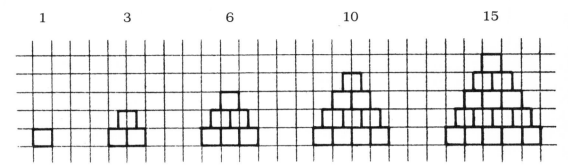

| 1 | 3 | 6 | 10 | 15 |

Triangular numbers, by the way, were discovered by the Pythagoreans. They considered the arrangement of 10 blocks or pebbles in a triangle and the number 10 associated with it as particularly important, and called it the "holy tetractys" (*tetra* is "four" in the Greek language, and 10 is the 4th triangular number).

While there is a short way of writing down square numbers, none is available for triangular numbers. Let us therefore *invent* one: let a triangular number be marked by a triangle, with a number at the bottom right corner giving its place in the series of such numbers. For instance,

$$\Delta_4$$

is the 4th triangular number, representing a pile with 4 blocks in the bottom row. We will call it, for short, "triangle-four." Since the pile representing this number contains 10 blocks, we may write

$$\Delta_4 = 10,$$

and you may read this, if you wish, as "triangle-four equals ten."

The first two triangular numbers are

$$\Delta_1 = 1 \quad \text{and} \quad \Delta_2 = 3.$$

Solution

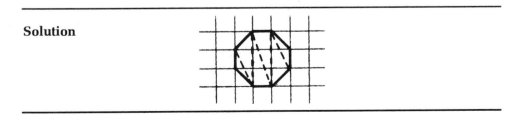

To get the next triangle we add a row of 3 blocks below the pile we already have

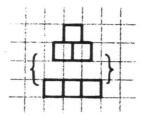 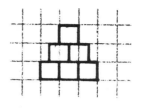

so we get

$$\Delta_3 = \Delta_2 + 3 = 6.$$

To find Δ_4 we add a row of 4 blocks below the existing pile

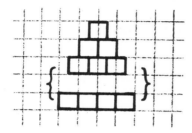 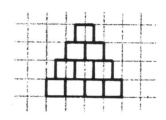

and get

$$\Delta_4 = \Delta_3 + 4 = 10.$$

This process can be continued as many times as one wishes. The *rule* it gives should by now be clear: any triangular number can be found by adding the small number with which it is marked (its "index," as such numbers are called in mathematics) to the triangular number coming *before* it. For instance,

$$\Delta_5 = 10 + 5 = 15$$
$$\Delta_6 = 15 + 6 = 21$$
$$\Delta_7 = 21 + 7 = 28$$
$$\Delta_8 = 28 + 8 = 36$$
$$\Delta_9 = 36 + 9 = 45$$

and so on.

This is a rather slow method: to reach Δ_9 one must carry out 8 additions. Is there a faster way? Yes indeed, as will now be shown.

For example, let's find Δ_6. We begin by drawing a pile of Δ_6 squares of our square-ruled paper, in such a way that the bottom and one side of the pile are both straight:

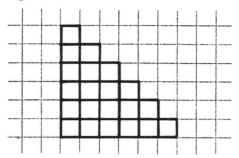

Next we add another such pile—but draw it upside down:

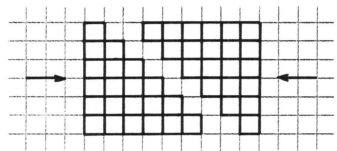

If one pushes the two figures together one gets a *rectangle*. The height of this rectangle is the same as that of one of the triangular piles—that is, 6 units. The width of the rectangle is one unit more than the width of a pile—in the present case, this means 7 units.

The number of squares contained in the rectangle is therefore

$$6 \times 7 = 42,$$

and this is equal to the number of squares in *two* equal piles:

$$\Delta_6 + \Delta_6 = 42.$$

To get Δ_6 we must take *half* this number—that is, we divide by 2:

$$\Delta_6 = \frac{42}{2} = 21.$$

This agrees with the result found earlier by simple addition. The same rule works for other numbers, too. To find Δ_7, multiply 7 by the number following it—which is 8—and then divide the result by 2:

$$\Delta_7 = \frac{7 \times (7+1)}{2} = \frac{7 \times 8}{2} = \frac{56}{2} = 28.$$

It makes no difference here if one first divides one of the numbers and only afterwards performs the multiplication. For instance, one could first divide 8 by 2 to get 4, then multiply by 7 to get 28: this is actually a simpler way, since one then deals with smaller numbers.

Next comes

$$\Delta_8 = \frac{8 \times (8+1)}{2} = \frac{8 \times 9}{2} = \frac{72}{2} = 36,$$

or, if you prefer to divide first and multiply last,

$$\Delta_8 = \frac{8' \times (8+1)}{2} = \frac{8}{2} \times 9 = 4 \times 9 = 36.$$

You may test Δ_9 yourself, if you wish.

All this can be written neatly in a *formula*:

$$\Delta_N = \frac{N \times (N+1)}{2}.$$

To use the formula, one replaces N with the appropriate number and calculates the result.

There is a story related to this formula, concerning a boy named Carl Friedrich (German for Charles Frederick) Gauss. He was born in Germany, to a poor bricklayer, in the year 1777—one year after the U.S. Declaration of Independence was signed.

From his early childhood Gauss was interested in numbers—in later years he used to say that he could count before he could talk. Once when he was three years old he was present when his father was calculating the payment to a group of workers in which he served as foreman. To everyone's surprise, little Carl told his father that there was a mistake in his sum—and when the calculation was checked this was indeed found to be the case.

When Gauss was ten years old his class was taught by a man who apparently did not believe in spending much of his time teaching. He gave the children a long exercise in addition—adding together the first 100 whole numbers:

$$1 + 2 + 3 + 4 + \ldots + 99 + 100.$$

The teacher figured that this would keep the children busy for the hour, adding up number after number on the little chalkboards used in schools in those days. He himself, of course, knew by the formula that the answer was

$$\Delta_{100} = \frac{100 \times 101}{2} = 50 \times 101 = 5050.$$

However, no sooner had the teacher completed giving the problem when Gauss wrote one number down on his board, slapped it down, and announced that he had finished.

Indeed, when finally all the children handed in their work, Gauss was the only one with the right answer. Later he explained how he did it. He arranged the numbers in pairs, working from both ends of the list:

$$1 + 100 = 101$$
$$2 + \ 99 = 101$$
$$3 + \ 98 = 101$$

and so on. The hundred numbers formed 50 pairs each of which equaled 101, so the answer had to be

$$50 \times 101 = 5050.$$

The teacher was sufficiently impressed to buy Gauss an advanced textbook on mathematics. Gauss later became a famous mathematician and scientist: he carried out the first exact measurements of the earth's magnetic attraction (a basic unit in magnetism is called the "gauss" in his honor), investigated the laws of probability, and derived many important results in mathematics.

11

The Sum of Squares

Suppose you had not read the preceding discussion but just saw the final result, that is, the formula

$$1 + 2 + 3 + \ldots + N = \frac{N \times (N+1)}{2}.$$

It would be hard to blame you if you felt somewhat suspicious about this result. On the left side of the equality sign, you might have said, stands a sum of whole numbers, which itself must also be a whole number. But on the right side there is a whole number divided by two, and you know well that such division often gives a number with a fractional part, *not* a whole number. Isn't it possible this could happen here?

There is a very good reason why it never happens. If division by 2 is to give a result with a fraction, the number being divided must be *odd*. However, $N \times (N+1)$ can never be odd: we get it by multiplying a number by the number following it, and of two such numbers, one is sure to be even (two odd numbers never follow each other!). So *either* N or $N+1$ can be divided by two, and the same is true for their product: the formula therefore never gives a fraction. There also exists a formula for the sum of squares:

$$1 + 4 + 9 + \ldots + N^2 = \frac{N \times (N+1) \times [(2 \times N) + 1]}{6}.$$

You could try it for a few numbers (including $N = 1$) and check it out, but the same doubt that was described before may still remain: how can one be sure that the number on top of the fraction on the right can always be evenly divided by 6?

For a number to be divisible by 6, it must be divisible by both 2 and 3. Furthermore, if the product of three numbers (the result of multiplying them together) is to be divisible exactly by 2 and by 3, then at least *one* of the numbers must be divisible by 2 and at least one (the same number or a different one) by 3. That much you probably know from your experience with numbers, although the exact mathematical proof is not so easy.

There is no difficulty in showing that the product

$$N \times (N+1) \times [(2 \times N) + 1]$$

can always be divided by 2, since we have just seen that one of the first two numbers in it *must* be even. To show that it is also divisible by 3, however, takes some more doing.

With regard to divisibility by 2, every whole number N is either odd or even. If N is *even*, it is exactly *twice* some other whole number M:

$$N = 2 \times M. \quad (N \text{ even})$$

If it is *odd*, it is larger by 1 than some even number and must therefore be of the form

$$N = (2 \times M) + 1. \quad (N \text{ odd})$$

Similarly, with respect to divisibility by 3, any whole number N must belong to one of *three* kinds of numbers. It may be *exactly* divisible by 3, and if so then a whole number M exists so that

$$N = 3 \times M.$$

It may be *larger* by *one* than such a number:

$$N = (3 \times M) + 1.$$

Or it may be larger by *two*:

$$N = (3 \times M) + 2.$$

If it is larger by *three*, it will have the form

$$N = (3 \times M) + 3,$$

and it is easily seen that it can be divided into three equal parts, each of which is $(M+1)$: we are therefore back to a number of the first kind. You can check a list of numbers in their natural order and see for yourself that the three kinds appear in a regular fashion, similar to the way in which odd and even numbers alternate regularly. For instance:

$$9 = 3 \times 3$$
$$10 = (3 \times 3) + 1$$
$$11 = (3 \times 3) + 2$$
$$12 = 3 \times 4$$
$$13 = (3 \times 4) + 1$$
$$14 = (3 \times 4) + 2$$

and so on.

Now let us look at the product

$$N \times (N+1) \times [(2 \times N) + 1].$$

If N is a number of the *first* kind, one of the three numbers multiplied here is certainly divisible by 3—namely, N itself.

If N is a number of the *third* kind, then $N+1$ is divisible by 3, for we have

$$N + 1 = [(3 \times M) + 2] + 1 = (3 \times M) + 3,$$

and it was already shown that this can be divided by 3.

Finally, if N is a number of the *second* kind, it can be written as

$$N = (3 \times M) + 1.$$

In that case, the last number of the product is

$$(2 \times N) + 1 = N + N + 1$$
$$= (3 \times M) + 1 + (3 \times M) + 1 + 1$$
$$= (3 \times M) + (3 \times M) + 3.$$

(If you have studied some algebra you ought to be able to perform this calculation much more quickly and neatly than is shown here.) In the last sum each number can be divided by 3 and the same is therefore true for the sum itself. Thus no matter to which of the three kinds of number N belongs, *one* of the numbers multiplied in the product is always divisible by 3 and

$$\frac{N \times (N+1) \times [(2 \times N) + 1]}{6}$$

is always a whole number.

The formula for the sum of third powers ("cubes") is

$$1 + 8 + 27 + \ldots + N^3 = \left[\frac{N \times (N+1)}{2} \right]^2,$$

where one first forms the number inside the large brackets and then squares it. Since the number being squared is Δ_N (and therefore whole) we need not worry about fractions here.

12

Magic Squares

Let us join 9 squares in a square block, like this:

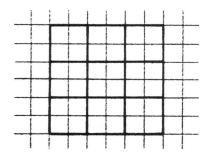

As in the problem with the prisoner, we shall call these squares *cells* for short. Because later some numbers will be written in these cells each of them will be two units wide.

Now here is the problem: Can the numbers from 1 to 9 be placed in these nine cells in such a way that we get the same sum of numbers . . .

. . . in any column   . . . in any row   . . . or in any diagonal?

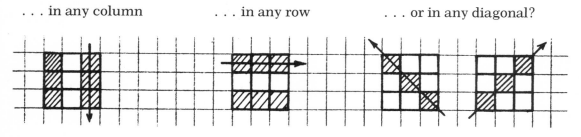

(A diagonal of a rectangle or a square is a line connecting opposite corners.)

It *is* possible.

Such squares are called *magic squares*, and many interesting things are known about them. Here we will show, step by step, how such a square can be put together—that is, how one finds in it the proper place for each number.

(1) THE FIRST STEP: THE SUM OF ANY COLUMN MUST BE THE SAME NUMBER. WHAT NUMBER IS IT?

We follow a trick already used before: when a number is not known, a *letter* is used in its place until its value is found. Here the unknown number is the sum of a column (or of a row, or a diagonal—they all should have the same sum) and will be marked with the letter N.

Now, what is the total sum of *all three* columns? There exist two ways of finding the answer.

First, because each column adds up to N, three columns should total

$$3 \times N.$$

Secondly, the three columns together contain all the numbers in the square, that is, all the numbers from 1 to 9. This sum should be

$$1 + 2 + 3 + 4 + 5 + 6 + 7 + 8 + 9 = \Delta_9.$$

One could find Δ_9 by adding all the numbers on the left; but it is faster to use the formula for triangular numbers:

$$\Delta_9 = \frac{9 \times (9+1)}{2} = \frac{9 \times 10}{2} = 9 \times 5 = 45.$$

Since both ways of adding up the three columns should lead to the same result, we must have

$$3 \times N = 45.$$

If three times N equals 45, then N must be one third of 45, or

$$N = \frac{45}{3} = 15.$$

> Therefore, each column, row,
> or diagonal adds up to 15.

(2) THE SECOND STEP: WHERE DOES ONE PUT THE NUMBER 9?

Let us check how many sums of three different numbers exist—out of the numbers available to us—that add up to 15 and contain the number 9. We have

$$9 + 1 + 5 = 15$$
$$9 + 2 + 4 = 15$$

and this is *all*: the sum

$$9 + 3 + 3 = 15$$

cannot be used, for every number may appear only once. Therefore the number 9 can appear (at most) in only *two* such sums.

Because of this, 9 cannot appear in the middle cell, for then it would have to belong to *four* different sums, marked by arrows in the drawing:

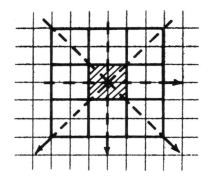

It cannot even be in a corner, for then it would belong to *three* different sums:

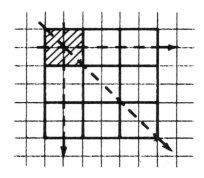

It *must* be on a side, where only two sums are involved:

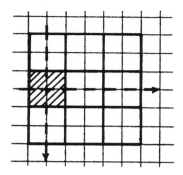

We must therefore have:

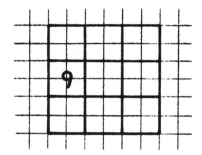

Of course, 9 could appear in any other side cell, but one can always bring it to the position drawn above by turning the square around:

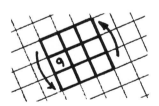

(3) THIRD STEP: WHERE DOES THE NUMBER 1 GO?

This is solved in much the same way as finding where to put the 9. The number 1 can only belong to *two* sums:

$$1 + 9 + 5 = 15$$
$$1 + 8 + 6 = 15.$$

The third possibility

$$1 + 7 + 7 = 15$$

contains the number 7 twice and therefore may not be used. Like the number 9, therefore, the number 1 cannot fit into the middle cell or a corner but must be placed on a side.

Note also that one of the sums containing the number 1 also has in it the number 9. This means that 1 and 9 are in the same row:

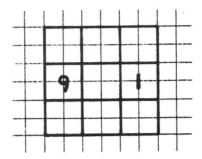

The third number in that sum is 5, so it can only fit in the middle:

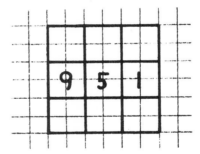

(4) THE REMAINING STEPS:

The other sum which includes 1 is

$$1 + 8 + 6 = 15.$$

Let us add it to the magic square (we could, if we wished, exchange 8 and 6):

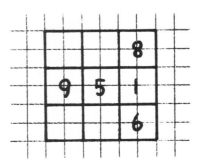

By noting that the diagonals also must add up to 15 two more numbers may be added:

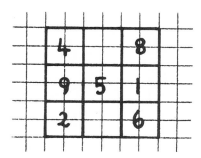

The remaining two numbers are now easily found, giving this as the final result:

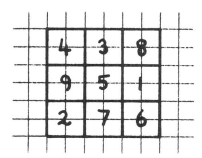

This is the only pattern possible, except for related patterns obtained by turning this one around and in addition perhaps exchanging left and right columns. It has been known for a very long time: the Chinese, who were the first to discover it, claim that its inventor was Emperor Yu of China, who lived around the year 2200 B.C. and who (they tell us) discovered the correct numbers from markings he saw on the shell of a turtle. In Europe, many centuries ago, people believed that this square indeed had magic powers, and they used it in charms worn to keep away devils and evil spirits.

The magic square described here has a total of 8 sums, each of them adding up to 15—3 in rows, 3 in columns, and 2 in diagonals. If we tried to list all possible groups of 3 different whole numbers between 1 and 9 having the sum 15, we would find just these 8 sums and no additional ones besides them. Keeping this in mind, see if you can discover the trick in the "game of 15":

On a table, face up, there are 9 slips of paper, marked with the numbers 1 to 9. Two players take turns, each time taking one slip from among those left on the table: the first player to possess 3 slips with numbers adding up to 15 is the winner. If, after all the slips have been picked up (one player then has 4 slips and the other one has 5), neither of the players has in his collection 3 slips adding up to 15, the game is declared to be a draw. Suppose you were asked to play this game—how would you choose the slips you picked up?

Hint: Remember the magic square! The answer is given on page 46.

13

Larger Magic Squares

Larger "magic squares" also exist. For example, here is a "four by four" magic square (four rows of four cells each).

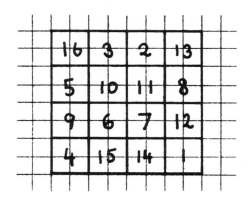

The "magic number" here is 34; one can prove that it *must* be 34 by a method similar to the one by which we showed that the "magic number" of the smaller square was 15. (The proof is given on page 49, but try first to arrive at it by yourself!) This particular square is famous for being included in the picture "Melencolia I" (I here is a Roman numeral), drawn by the great German artist Albrecht Dürer in the year 1514: it has the date of the picture cleverly hidden in the two middle cells of the bottom row. (The sound of German *ü* is pronounced like "ee" but with lips puckered, as if you intended to say "oo".)

An interesting problem is connected with this particular square. So far we have discussed magic *squares*, but there can also be other

"Melencolia I"

MELANCHOLIA. B-6,547. Albrecht Durer. National Gallery of Art, Washington. Rosenwald Collection. Reproduced by permission.

"magic" shapes. Take for instance a 4-by-4 square with the bottom right-hand corner missing, as shown in the drawing. There are 15 cells in this shape. Can one place in them the numbers from 1 to 15 so that the sum of every row, column, or diagonal is the same (namely, 30)? It *is* possible: can you do it?

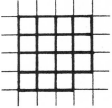

Hint: the solution is closely related to the magic square of Albrecht Dürer. Although the problem appears to be quite difficult, there is a trick by which the right answer can be found almost immediately. It is explained on page 52.

Among the many people who enjoyed magic squares in the past was Benjamin Franklin. Franklin made up magic squares to pass the time during long and boring sessions of the Pennsylvania General Assembly—the law-makers of colonial Pennsylvania—where he was appointed clerk in 1736. He later wrote:

> In my younger days, having once had some leisure, (which I still think I might have employed more usefully) I had amused myself in making these kind of magic squares, and, at length, had acquired such a knack at it, that I could fill the cells of any magic square, of a reasonable size, with a series of numbers as fast as I could write them, disposed in such a manner, as that the sums of every row, horizontal, perpendicular or diagonal, should be equal. . .

Solution

The "game of 15":

Imagine yourself as one of the players. Draw a "three-by-three" magic square, with its nine numbers, and as the numbered slips are chosen, draw on the square an O around each number taken by you and cross out with an X each number picked by the other player.

At any given time, the unmarked numbers on the magic square represent the slips that still remain on the table, and whenever it is your turn you will place an O around one of the numbers. Choose the number on the square before picking up the slip that carries the number. If you can, choose it in such a way that the O you mark gives you three Os in a straight line (row, column, or diagonal), for then the numbers on the slips to which they "belong" add up to 15. If this is not possible, place your O in a way that will help you get three-in-a-line later on.

By now you have surely guessed the trick: if you mark the chosen numbers with Os and Xs on the magic square, playing the "game of 15" becomes exactly the same as playing tic-tac-toe. If you are skilled in tic-tac-toe, you should have no trouble placing your Os so that either you will be the winner or the game will end in a draw.

Of course, Franklin must have used some special method to produce magic squares with such speed. Several such methods exist, and one of them is described in what follows. It works for any magic square with an *odd* number of cells on each side.

The method, by the way, bears the name of a Frenchman, Antoine de la Loubère, who learned it while visiting the Kingdom of Siam (now Thailand) in the years 1687 and 1688. It had been known for many years before in Siam, India, and neighboring countries.

200	217	232	249	8	25	40	57	72	89	104	121	136	153	168	181
58	39	26	7	250	231	218	199	186	167	154	135	122	103	90	71
198	219	230	251	6	27	38	59	70	91	102	123	134	155	166	187
60	37	28	5	252	229	220	197	188	165	156	133	124	101	92	69
201	216	233	248	9	24	41	56	73	88	105	120	137	152	169	184
55	42	23	10	247	234	215	202	183	170	151	138	119	106	87	74
203	214	235	246	11	22	43	54	75	86	107	118	139	150	171	182
53	44	21	12	245	236	213	204	181	172	149	140	117	108	85	76
205	212	237	244	13	20	45	52	77	84	109	116	141	148	173	180
51	46	19	14	243	238	241	206	179	174	147	142	115	110	83	78
207	210	239	242	15	18	47	50	79	82	111	114	143	146	175	178
49	48	17	16	241	240	209	208	177	176	145	144	113	112	81	80
196	221	228	253	4	29	36	61	68	93	100	125	132	157	164	189
62	35	30	3	254	227	222	195	190	163	158	131	126	99	94	67
194	223	226	255	2	31	34	63	66	95	98	127	130	159	162	191
64	33	32	1	256	225	224	193	192	161	160	129	128	97	96	65

One of Ben Franklin's "Magic Squares"

To show how the method works, we will use it to produce a five-by-five magic square. Let us begin by drawing a five-by-five square and then adding three other squares of the same size to the right of it and above it, as shown in the illustration.

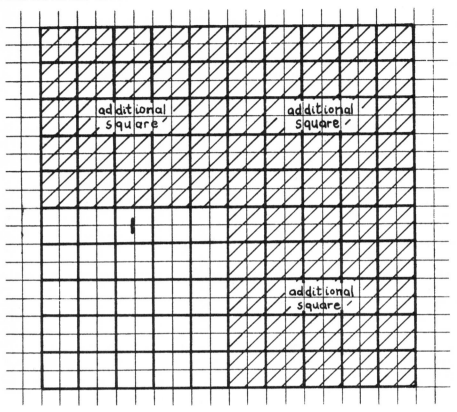

One begins by placing the number 1 in the top middle cell (as shown in the drawing) and then proceeds to add the numbers following 1, in their natural order, using *three rules*.

Rule 1: Always try to advance diagonally upward to the right, like this:

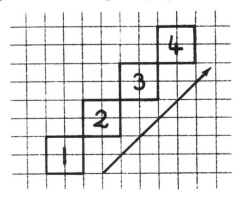

Rule 2: Sooner or later, as one advances in the diagonal direction, one may arrive outside the main magic square, in some cell belonging to one of the *additional* squares. The appropriate number then goes into the cell having the same position in the *main* square.

For instance, by the first rule the number 2 would go into the cell to the right of the bottom middle cell in one of the additional squares. It is instead placed—by the second rule—to the right of the bottom middle cell in the *main* square.

Where the number 2 would go according to the first rule →

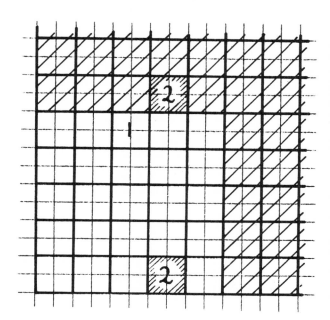

Where the number 2 actually goes, according to the second rule →

Rule 3: Occasionally, as one follows either of these rules, one finds that the next cell to be filled already has a number in it. One then drops down and fills the cell below the last one (*not* below the one that's blocking the way).

Solution

The "magic number":

If M is the "magic number," then

$$4 \times M = \Delta_{16} = \tfrac{1}{2} \times 16 \times 17 = 8 \times 17 = 136.$$

Therefore M is one quarter of 136: dividing 136 by 4 then gives $M = 34$.

The next drawing is the complete square. We also show there (with small numbers in the shaded cells) where some of the numbers would have been placed if the second rule were not used. Note that by the second rule, the number 16 should occupy the space already taken by 11: by the third rule one must therefore descend one space and place it below 15, as shown.

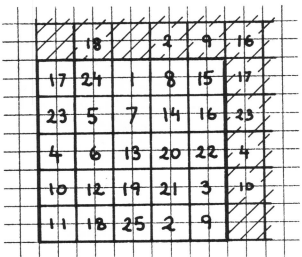

The seven-by-seven square derived by the same method is given below:

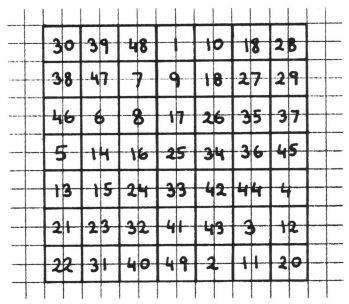

If you have followed the construction of these squares, you can easily handle bigger ones. Soon you will find it no longer necessary to draw the additional squares and will develop just as much of a "knack" at it as Ben Franklin himself had!

14

The Size of the Squares

How long is each square on our ruled paper?

Before we can answer this, we must know more about measurement of length. In the United States length is usually measured in *inches*. This rectangle is one inch long:

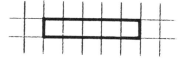

This one—two inches:

Three inches:

To measure length, a *ruler* is used. The type of ruler used in schools and offices is usually 12 inches long—this length is called *one foot*—and there are generally marks along its edge, dividing it into inches and into parts of inches—halves, quarters, eighths, and sixteenths.

The squares on square-ruled paper are usually a *fifth* or a *quarter* of an inch long, that is, a row of 5 or 4 squares has the length of one inch. The drawings in this chapter are all made on paper with five squares to the inch.

But what *is* an inch?

The word *inch* comes from Latin, the language of the Romans, who 2000 years ago already had a complete system of measurements. (Our names for the months of the year also come from the Romans.) In Latin, *uncia* (pronounced "oonkyah") means "one twelfth"—in this case, one twelfth of the length of one foot. In Old English it came to be spelled *ynce* (pronounced "incha"), and eventually it acquired its modern form.

Curiously, the word *ounce* also comes from the same source (but via the French *unce* or *once*). Our pound started as a Roman unit of weight called the *libra*, a name still echoed by the abbreviation "lb" used for the pound. This Roman "pound" however, was only about ¾ the size of the modern pound. The Romans used the word *uncia* to describe one twelfth of the weight of one pound, as well as one twelfth of the length of one foot. The French word for the weight unit gradually became the English *ounce*. The pound was later changed to contain not 12 but 16 ounces, but each ounce remains almost the same as the *uncia* of the Romans. The old "troy pound" used by jewelers in weighing gold and silver still has only 12 ounces, though these "troy ounces" (which were first used in the French city of Troyes) are slightly larger than the regular ones.

Now, back to the measurement of length. The inch, as we have seen, is related to the foot, which—as the name suggests—is about equal to the length of the foot of a grown person. Hundreds of years ago, at places where people did not have a marked length which they used as a standard foot of length, the feet of actual people were indeed used for measurement.

Solution

The magic nonsquare:

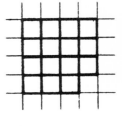

If you increase every number in a magic square by the same amount (say, by 3) the result is still "magic"—that is, the sum of each row, column, or diagonal is still equal. Of course, the smallest number in the square is then no longer 1 (if you added 3, it would be 4).

Similarly, if you take Dürer's square and subtract 1 from each number in it (making the largest number 15), the result is still "magic," but the smallest number is now zero, appearing in the bottom right corner. Since the zero makes no difference in any of the sums in which it appears, we get the same sums even if we cut that corner out—leaving exactly the "magic shape" we wanted! The original "magic number" was 34. The new one is 30 because in subtracting 1 from each number we are also subtracting 4 from each sum.

Here is one prescription, said to be over 400 years old:

To find the length of a rod in a right and lawful way, and according to scientific usage, you shall do as follows. Stand at the door of a church on a Sunday and bid sixteen men to stop, tall ones and small ones, as they happen to pass out when the service is finished: then make them put their left feet one behind the other, and the length thus obtained shall be a right and lawful rod to measure and survey the land with, and the sixteenth part of it shall be a right and lawful foot.

Thus, to make sure that the length "one foot" was the length of an *average* foot, one measured not just *one* person's foot but 16 feet placed one behind the other, each belonging to a different person! A few of these people would have large feet, others would have small feet, but with 16 different persons this would usually even out—more or less. The total length was called a *rod* and the sixteenth part of it was "one foot."

In later times, of course, the governments of both Great Britain and the United States had a "standard" ruler for measuring the exact length of a foot, (actually, the length marked was one yard, which equals 3 feet). All other rulers and measurements had to agree with this "standard."

Other old measures based on the human body include the *hand* and the *span*. A "hand" is 4 inches long and is (more or less) the width of the adult hand; it is still used in measuring the height of horses. A "span" is the distance between thumb and little finger when a hand (the "average" hand) is stretched as wide as possible: it is taken as about 9 inches. The Bible mentions the *cubit*, the distance from the elbow to the end of one's stretched fingertips: it is approximately 18 inches, or a foot and a half. The giant Goliath, whom David killed with his slingshot, is reported in the Bible to have had a height of "six cubits and a span" (I Samuel xvii. 4). With the preceding information you should have no trouble in converting this into feet and inches!

15

Making Rulers

By means of square-ruled paper you can make your own rulers. Suppose the squares are one-fifth of an inch wide: then by cutting a strip of this paper you can make a ruler on which every inch is divided into five equal parts:

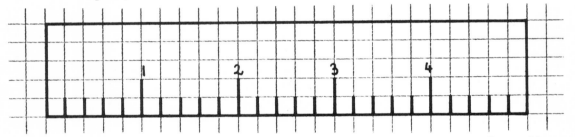

Can we make a ruler divided in a different way—say with markings one quarter of an inch apart? Indeed we can. The picture which follows shows how this is done—or more exactly, how an inch is divided into 4 equal parts:

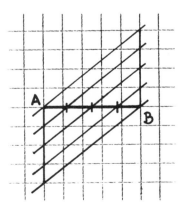

The line from A to B is one inch long. In order to divide it we use the square ruling to draw a bundle of 5 lines which are all parallel—that is, they all have the same direction—so that the distance between any two neighboring lines in the bundle is the same. Such a bundle cuts any part of a straight line which crosses its way into 4 equal parts, and this is exactly what it does to the marked inch.

To draw the bundle we add two lines perpendicular to the line which we want to divide, one at each end—one of them pointing up, the other pointing down. On each line we mark off some chosen length 4 times (which is easy to do with the square ruling). Connecting the marks then gives the required parallel lines. Note that the two outermost lines of the bundle really need not be drawn, for only the three inner lines are required for drawing the dividing marks!

The same method can be used for dividing the inch (or any other length) into any number of equal parts. For three parts we have

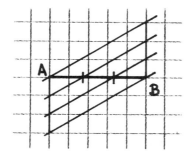

A third of an inch used to be called a *barleycorn*—supposedly, the length of a grain of barley. No one measures length in barleycorns any more, except perhaps shoe salesmen, for the length of a shoe increases by about ⅓ inch whenever its "size" increases by one—that is, a size-8 shoe is one barleycorn longer than a size-7 shoe, and so forth.

Many other old measures exist. The *rod* (or "pole") was usually taken as 16½ feet—not 16, as given in the prescription for measuring one foot. Four rods made a *chain* of 66 feet—the length of a chain used for measuring land, introduced by the English mathematician Edmund Gunter around 1620. Ten chains made a *furlong* and 8 furlongs a *mile*. Of these, only the mile is still being used, although land areas are still measured in acres, an *acre* being the area of a rectangle one furlong long and one chain wide.

16

Meters and Centimeters

Very few countries outside the United States still measure length in inches and feet. Instead they use the "metric system" in which the basic unit of length is the *meter*. One meter is slightly less than 40 inches, so you may think of the meter as a "long yard" (1 yard = 3 feet).

Additional units of length exist in the metric system and are used for measuring distances much larger or much smaller than the meter. They are all related to the meter and this relation always involves numbers like 10, 100, or 1000: no odd-sized numbers like 16½ or 5280 (the number of feet in a mile) ever appear.

Here are some of these units:

A *decimeter* is one-tenth of a meter and comes very close to 4 inches:

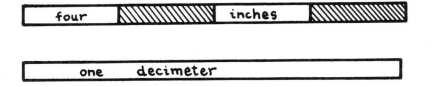

A *centimeter* (written cm for short) is one-hundredth of a meter. There are 10 centimeters in each decimeter:

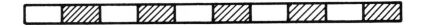

In Europe square-ruled paper generally has squares half a centimeter wide. It looks almost exactly the same as paper ruled with 5 squares to the inch. Only by using a very accurate ruler, or by trying to match the edges of two sheets, one of each kind, can one tell apart the two types of ruling.

One-tenth of a centimeter is a *millimeter* (mm for short) and there are 1000 millimeters in a meter. Paper with square rulings spaced one millimeter apart looks like this:

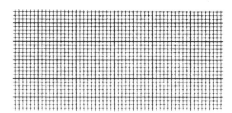

Decimeters, centimeters, and millimeters are all used for measuring small distances. Large distances are measured in *kilometers* (km for short), each of which equals 1000 meters. One kilometer is somewhat longer than half a mile—in fact, 8 kilometers come very close to 5 miles, so that a speed limit of 50 miles per hour in the U.S. would translate to 80 kilometers per hour elsewhere. A friend of mine who visited India some years ago reported seeing there a road sign giving the speed limit as 24 km per hour: he wondered why such a non-round number had been chosen and then realized that the country had just changed over to the metric system and that the sign he saw probably replaced an earlier one which had read "15 miles per hour."

The meter (its name is from the Greek word for "measure") was introduced in France in the year 1791, to replace the old French system of units. That system had included the French "foot," supposedly the length of the foot of the famous King Charles the Great ("Charlemagne") who lived 1000 years earlier: it was slightly longer than the English foot (also used in the U.S.). Perhaps the difference had to do with the "greatness" of Charles.

The meter, on the other hand, was based on a quite different standard—the size of the Earth: it was supposed to be equal to one part in 10 million of the distance from the pole of the Earth to the equator. The French actually measured a distance on Earth (it is enough to measure only *part* of the distance from the pole to the equator, if appropriate astronomical observations are made at the end points) and from this found what they believed was the proper length of one meter. This length was then marked by two scratches on a metal rod, placed one meter apart, and the rod became the "standard meter" with which all rulers and measuring instruments were compared.

By the way, in addition to the meter there exists one other unit of length related to the size of the Earth, namely the *nautical mile* ("nautical" means having to do with ships and sailing). The distance from pole to equator is also a quarter of a "great circle" passing through both poles of the Earth, and circles (as well as angles) have been traditionally divided into *degrees*—360 degrees in a full circle. Each degree in its turn is divided into 60 *minutes* and each minute into 60 *seconds* (to avoid confusion with units of time, these are often called "minutes of arc" and "seconds of arc").

One nautical mile is defined as equal to one minute of arc on the "great circle" just described. How far is this in meters? The answer, together with some more facts about nautical miles, can be found on page 61.

But back to the "meter." Later measurements found that an error had occurred and that the distance between the scratches differed slightly from one ten-millionth of the distance between pole and equator, but it was too late to change the units. Nowadays there exist measurements which require more accuracy than can be obtained by measuring the size of the Earth, or even the distance between two fine scratches on a metal bar. Instead, the length of the meter is nowadays related in a complicated way to the atomic properties of matter (like the wavelength of light emitted by certain atoms)—properties that can be measured very precisely and (supposedly) never change.

Will the United States ever switch to measure in meters? Probably, yes—perhaps even fairly soon. The British (who use the "metre" spelling) started doing so in 1965, stretching out the change over many years, and the U.S. now remains the last large country still measuring distances in feet and miles. Scientists everywhere have been using the metric system for many years. And you *might* say that (in a way) the U.S. already has switched to the metric system: on April 5, 1893, a law passed by Congress ordered that all units of length in this country be defined by the number of centimeters they contain and not in terms of any "standard" foot or yard!

17

Games

There are many games that can be played on square-ruled graph paper. Two of them are described here, each intended for two players:

Completing the Squares

This game is said to have been invented by students who used graph paper in their work at the French Polytechnic School, a famous college for engineers.

Draw a square 8 units wide on square-ruled paper (bigger squares may be used, but then the game takes longer). The square will contain $8^2 = 64$ small squares which, as before, will be called "cells."

The players now take turns tracing with a pencil *one* side of any cell inside the big square: since square-ruled paper is usually printed in light blue, the traced sides will be plainly visible. Any side which has already been traced may not be traced again; this includes the borders of the big square, which were already traced when the square was drawn.

The aim of the game is to *enclose cells from all four sides*. A player who manages to enclose a cell (by drawing its fourth side) receives a point, and at the end of the game whoever has more points wins. For instance:

if a player traces *this* line . . .   . . . he gets a point for enclosing *this* cell.

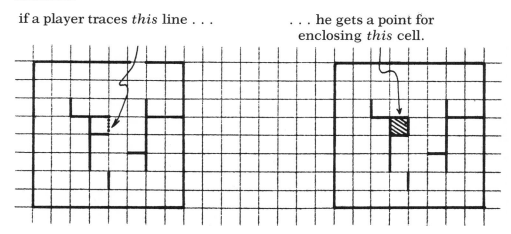

Next to a border only three sides have to be traced to enclose a cell; and in the corners, only two sides: the player who traces the side of a corner cell thus gives his opponent a chance to score a point at the next turn. Players who have scored a point get a free extra turn; and if they score on that, another free turn, and so forth—they continue tracing lines until they stop scoring.

At the beginning of the game it is easy to trace lines in a way that will not allow the opponent to win any points. But as the area is filled, this becomes more difficult; and near the end one often finds that *any* line drawn will enable the opponent to score. The problem then is which move will give away the *fewest* cells!

Battleship

Back in the days when wars at sea were fought with big guns rather than with missiles, a *battleship* was a large warship protected by thick steel armor and carrying guns of great size and range. The name itself comes from the earlier days of sailing ships, when a common way of fighting a battle at sea was for the biggest ships to line up one behind the other so that their guns (poking out from their sides) were all pointed together at the enemy. This was called the "line of battle" and the ships were accordingly called "line-of-battle-ships" or "battleships" for short (another term for them was "ships of the line").

A smaller type of warship was called a *cruiser* (a few of these are still in use as this is written), a still smaller one a *destroyer*. Ships of any of these kinds usually fired their guns in groups, with several guns firing at once at the same target: this was called a *salvo*.

Each of the two players taking part in the game receives a sheet of square-ruled paper. On it each draws two squares 10 units wide—one for marking the position of the player's own fleet of ships and the other representing the area where his opponent's fleet is hidden. The squares are labeled as shown in the drawing that follows:

THE ENEMY'S FLEET   MY OWN FLEET

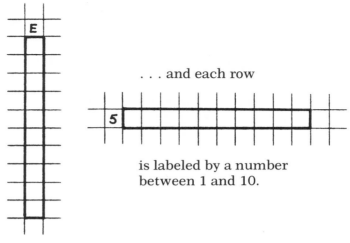

In this way, each column is labeled by one of the first 10 letters of the alphabet . . .

. . . and each row

is labeled by a number between 1 and 10.

Solution

The nautical mile:

A "great circle" contains 40,000,000 meters and also 360 × 60 = 21,600 minutes of arc. Dividing one number by the other gives the length of the nautical mile as 1852 meters, to the nearest meter.

The speed of ships is traditionally measured in *knots*, where one "knot" equals one nautical mile per hour. The reason for this odd name goes back to the time of sailing ships, when a ship's speed was measured with the help of a long rope in which knots were tied a fixed distance apart. The beginning of the knotted part was marked by a piece of cloth, and some distance beyond the cloth, at the end of the rope, a wooden float or "log" was tied.

In order to begin measuring the speed of a ship, the log was tossed overboard behind the ship: as the ship moved away from it, the rope was gradually pulled out. The moment the piece of cloth passed the end of the ship, a sandglass ("hourglass") timed to empty in 28 seconds was turned upside down, and the number of knots which passed overboard before the sand ran out was counted: clearly, the faster the ship moved, the more of the rope and the more of the knots went by. The distances between the knots and the length of time during which they were counted were all chosen in such a way that the number of knots counted was the same as the speed in nautical miles per hour.

The results of all such measurements were noted down in a "log-book" and from them the captain could carry out a calculation ("dead reckoning") giving the expected position of his ship. The name "log" is even now still used for the instrument by which a ship's speed is measured.

Using these labels, each cell inside one of the big squares can be identified by a combination of a letter and a number, giving together the column and row to which it belongs. For instance, cell A1 is the one in the top left corner, belonging to column A and to row 1. In the same way the remaining three corners are J1, J10, and A10.

Each player now places on his (or her) own square (without the opponent seeing it) a "fleet" of four "ships":

ONE BATTLESHIP   ONE CRUISER   TWO DESTROYERS

These may be placed anywhere—horizontally (along a row), vertically (along a column), or diagonally. If diagonally, they may appear like this:

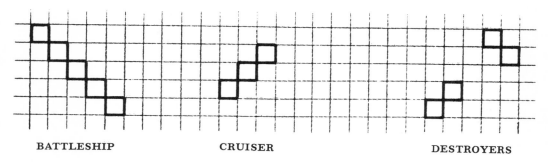

BATTLESHIP   CRUISER   DESTROYERS

No two ships may touch each other, even corner to corner. For a sample game, "my own fleet" might be placed as shown at the right in the illustration below.

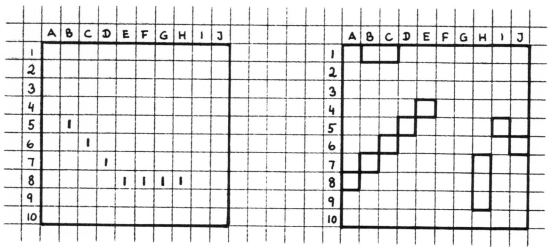

THE ENEMY'S FLEET   MY OWN FLEET

The two fleets now take turns shooting at each other, in salvos of 7 shots each. The aim is to hit and "sink" each other's ships, a ship being considered "sunk" if *all* its cells have been hit. For instance, a player may begin by announcing that the first salvo is aimed at the following cells:

$$B5, C6, D7, E8, F8, G8, H8$$

As the first salvo is fired, the player doing the shooting marks each of these cells on the square labeled "the enemy's fleet" by the number 1 (as shown in the drawing). The other player meanwhile marks the same squares on the map of his or her "own fleet" and then announces which ships were hit and how many times, but *not which shots* have scored the hits. This is what makes the game tricky: if an attacker is told, for instance, that the enemy battleship has been hit twice, he or she must still guess which were the 2 shots that scored, in order to direct the next salvo toward the other 3 cells of the battleship.

(It so happens, however, that if the salvo which has scored two hits on the battleship is the one in our drawing of the enemy fleet, no guessing is needed—it is possible to figure out *exactly* where the rest of the battleship is hidden. Can you do so? The answer is on page 66.)

It is then the other player's turn to shoot, and after that the first player fires another salvo, marking each target cell with a 2. If any player sinks a ship and can figure out which squares it had occupied, he or she can mark around it a "safe" area where no other enemy ship could have been placed, because the two ships would have touched. Each player also marks the opponent's shots on the square labeled "my own fleet," to prevent any later disagreement about the placing of shots.

A player whose battleship is "sunk" loses 3 shots and thereafter fires only 4 shots in each salvo (for this reason players often try to seek out the enemy's battleship first). Similarly, a player loses 2 shots with the sinking of his cruiser and one shot for each destroyer. The player who manages to sink all ships of the opposing fleet is the winner.

Different versions of the game exist. In one of these, each opponent fires only one shot at each turn, regardless of the number of ships left in his fleet. The fleet in this case consists of 4 "battleships," each having 4 squares; these squares may be arranged not only along rows, columns, and diagonals but also in a square 2 units in width (an odd shape for a ship, though about 100 years ago Russia *did* use *round* warships). Another version, played on a 9-by-9 square, has one battleship of 4 squares, 2 cruisers of 3 squares, 3 destroyers of 2 squares, and 4 submarines of one square each. Each salvo includes 3 shots, no diagonal ships or square battleships are allowed, and a ship may touch the side of the big square along at most one side of one square.

18

Coordinates

In the game of "Battleship," every cell was labeled by a letter and a number:

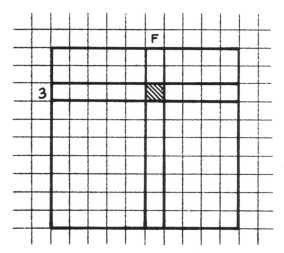

Maps of cities are often divided into cells in a similar way, as a help in finding the location of streets. Such maps usually have lists of streets printed along the margins or on the reverse side of the paper, giving for each street the cell (or cells) in which it may be found. If one reads there, for instance,

Lakeside Drive . . . D 14,

one knows that Lakeside Drive is to be found somewhere in cell D 14, which saves the trouble of looking for it all over the map.

Sometimes, however, it is not enough to say in which square on a map a place may be found: it becomes necessary to pinpoint its position as accurately as possible. A surveyor wants to know exactly where in a field a boundary marker is buried; an army officer wants to note down the exact location of an enemy outpost in the middle of a forest or desert. How can they use numbers to describe an exact point on their maps?

The scheme used in the game of "Battleship" and in city maps can still be used here, but it must be changed slightly, as follows.

First, we must discard the alphabet and use numbers for *both* labeled sides. Of course, with two sets of numbers serving as labels, one must be careful to give them different names and not mix them up.

Next, let the numbers mark not the squares but the division lines: each marked side then takes on the appearance of a small ruler. By general custom all markings begin at the *lower left-hand corner*: they proceed to the right along the bottom side of the map, which will be called the *x axis*, and upward along the left edge, which will be called the *y axis*. The markings on the edges then appear as in the drawing shown here:

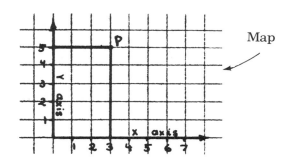

With these markings we cannot yet label *every* point on the map—but it *is* possible to do so for corner points at which two lines meet, such as the point P in the drawing. Using the lines of the ruling, we draw from P perpendicular lines to the two axes, as shown, and note the marked distances at which these lines meet the axes. The distance marked off on the *x* axis will be called *the x coordinate of P* while the distance on the *y* axis will be *the y coordinate of P* (the word is pronounced "co-ordinate"). The older names "abscissa" and "ordinate" for the two distances are still sometimes used, while scientists like to shorten everything and talk simply about "the *x* and the *y* of P" (so will we, very soon). In particular, for the example shown in the drawing, one can say that:

> The *x* coordinate of the point P is 3.
>
> The *y* coordinate of the point P is 5.

This could also be put more briefly:

> The coordinates of the point P are
>
> $x = 3, y = 5.$

Either way we say it, the meaning is that the point is marked by two numbers, 3 and 5. Or we could simply say:

> P is the point $(3,5)$.

Here we leave out most of the words, since only the numbers matter. This is how mathematicians would write it, with the understanding that the x coordinate is always the *first* number listed. This is important to note, because $(3,5)$ and $(5,3)$ are two quite different points, as the following drawing clearly shows:

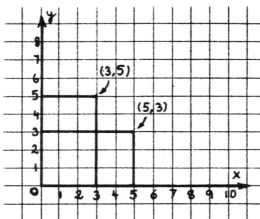

Solution

Sinking the battleship:

Among the shots fired, four are along a diagonal line and four are in a single row, and the fact that the battleship received more than one hit tells us that it could only lie along the diagonal or along that row. If it followed the diagonal, at least 3 of its cells would have been hit, because a battleship is 5 cells long; since that did not happen, it must lie along the row.

Four shots reached that row, but only two of them scored hits: the rest of the battleship must therefore be either to the right or to the left of these four. On the right side only two cells are available, which is not enough to contain the rest of the battleship: therefore the remaining part of the battleship must be on the left, in cells B8, C8, and D8.

19

Coordinates for Any Point

Up till now the "rulers" drawn along the axes were only marked in *whole* numbers; but if fractions are allowed, the in-between points can easily be marked as well. For instance, on either axis one might mark the point halfway between 5 and 6 as 5½ (or 5.5 in decimal fractions). Indeed, *any* point on an axis can be identified by a number giving its distance from the "zero mark" in the bottom left-hand corner—and this number can be whole, fractional (like 2/11 or 3¼ or 5.65), or even irrational!

The zero mark, by the way, has coordinates (0,0) and is called "the origin"—that is, "the starting place."

Now that in-between points on the axes are labeled, one can easily label any point on the map by its two coordinates. As before, one draws from the point lines perpendicular to the axes, and the numbers associated with the points at which these perpendicular lines meet the x axis and the y axis are the x and y coordinates belonging to the point. Actually, of course, no lines have to be *marked* on the map—it is enough to lay down a ruler along the directions of these lines and to note where it meets the axes.

Here are some examples:

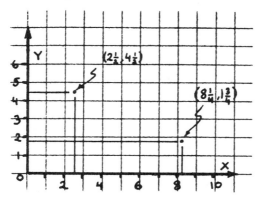

There remains one drawback to our method of labeling points: we can extend it as far as we please to the right and upward, but in the opposite directions our way is blocked by the axes. This restriction is removed if we continue the labeling of the axes on *both* sides of the origin, using negative numbers as shown in the drawing:

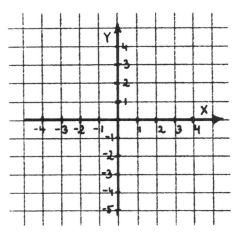

This continuation allows labels to be given to points *anywhere*—even below the origin and to the left of it, something which was not possible before. Some examples are shown in the drawing which follows: you are invited to check each of them to make sure that it is indeed labeled correctly.

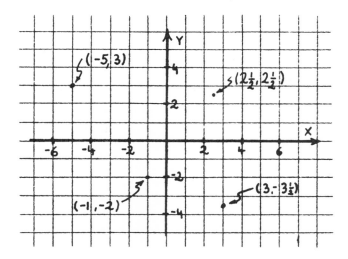

The system of coordinates described so far is the one which mathematicians prefer, and some of its uses will be described later on. You may hear it called the "Cartesian system" after the Frenchman René Descartes ("day-cart") who first described it in 1637. Survey and military maps generally arrange matters in such a way that only positive numbers appear as coordinates, and the finer division of the axes is expressed by decimal fractions or in a way that is equivalent to the use of such fractions.

Maps of large areas—say, of all of the United States—require a somewhat different system, made necessary by the fact that the Earth is not flat but ball-shaped: but again every point is marked by two numbers, called *longitude* and *latitude*. A similar system is used by astronomers to mark the place of a star in the sky, only they call the two numbers *right ascension* and *declination*.

Our coordinate system also resembles the labeling system for streets used in many cities. Two roads serve as axes—for instance, Main Street extending in the east-west direction and Central Avenue going north-south. Parallel to Main Street (that is, going in the same direction) we have, on the north side, First Street North, Second Street North (usually written 2nd Street N. for short) and so forth. Similarly, on the southern side one finds First Street South (or "1st Street S."), 2nd Street S., and so on, as shown here:

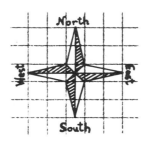

Likewise there might exist a First Avenue West ("1st Ave. W.") and a First Avenue East, and other avenues marked by larger numbers.

The practice of having two straight main streets perpendicular to each other and meeting in the middle of a city is actually very old, dating back at least to the early cities built by Greeks and Romans over 2000 years ago. Such streets divide the city into four parts, which were accordingly called *quarters*. For instance, the old walled city of Jerusalem, which is more or less rectangle-shaped, is divided in this manner into four quarters, one to each corner: the Moslem (Arab) Quarter, the Christian Quarter, the Jewish Quarter, and the Armenian Quarter (Armenians are a Near-Eastern Christian nation). Indeed, the word "quarter" is sometimes used in English and in other languages to denote simply a part of a city. The old section of New Orleans, dating back to the time when the city belonged to France, is still called the "French Quarter."

20

More Games

Race Track

In this game, two or more players follow a "race track" drawn on square-ruled graph paper, advancing marks which represent their "racing cars." Just as it happens in a real race, here too some time is needed for a car to build up speed, to slow down, or to change direction; and a car entering a curve with too much speed risks hitting the edge and "getting wrecked." The game seems to have been invented in Europe, from where it was brought to the U.S. by Jurg Nievergelt of the University of Illinois. A detailed description of it was given in Martin Gardner's column "Mathematical Games" in the January 1973 issue of *Scientific American*.

On a sheet of square-ruled paper a "race track" is drawn, with a "starting line" along one of the lines of the ruling (wide enough to accommodate all starting cars: typically 2 to 4) and a "finish line" at the end (see the drawing on page 72). Several curves should be included in the track to add interest to the game, and a new and different track may be drawn for each game. Each player uses a pen or a pencil of a different color to mark the position of his or her car, which must always be located on one of the intersections of the ruling.

All cars begin at the starting line, and players then take turns (in an order determined by the toss of a coin or in a similar way) advancing them according to rules described below. With every advance of a car, its new position is marked on the sheet in the appropriate color and is then connected by a straight line of the same color to its previous position. As in real car racing, the first player to cross the finish line wins.

To explain the rules by which the cars advance it is best to imagine a system of (x,y) coordinates covering the sheet, with one unit of length along each of the axes corresponding to the width of one square (see the drawing). In each "step" of the game, each coordinate (x and y) changes

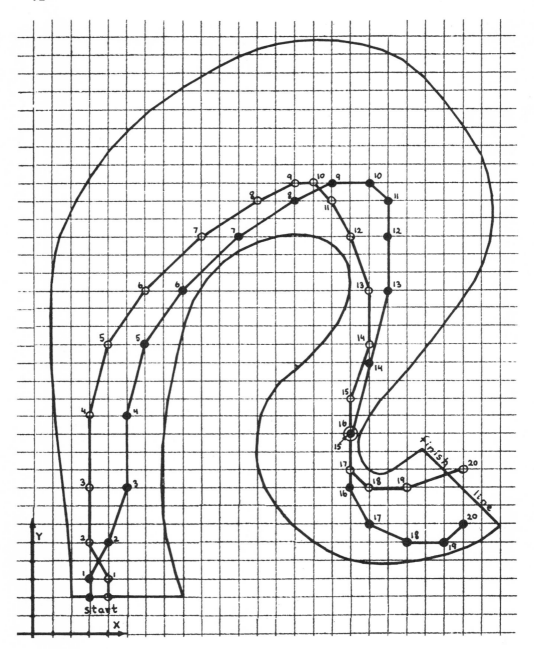

by a whole number which may be positive (indicating an increase), zero (indicating no change), or negative (indicating a decrease).

The change in x allowed by the rules, in a given step, always depends on the change in x during the *preceding* step: it may either be the same, or larger by 1, or smaller by 1. For instance, if the previous change in x was $+2$ (meaning the car moved 2 units to the right), the next must be either $+2$, or $+3$, or $+1$ (moving 1, 2, or 3 units to the right). A similar rule holds for the change in y. For the first step (off the starting line) one assumes that the "previous step" involved *no* change in either x or y, and therefore the first change in either x or y must be either zero, or $+1$, or -1.

All the steps of the game drawn here follow these rules, as you may check for yourself.

Additional rules must also be followed. Neither the marked points nor any parts of the straight lines connecting them are allowed to leave the race track, and any player whose car crosses the boundary loses. Also, no two cars may occupy the same spot at the same time (that would be a collision!), although their tracks may cross and a car may be placed at a point where another car had been earlier.

If the first car to cross the finish line does so after, say, 24 steps, it is only fair to allow any car that has only completed 23 steps to take one more turn, so that all cars will have completed the same number of moves. After that, if two or more cars have passed the finish line, the car that has gone the greatest distance past the line is the winner.

A short sample "race" is shown in the drawing, with the positions of the two competing cars labeled by the numbers of the steps in which they were reached. At first, the car marked by the heavy black dots pulls ahead, but it overshoots the second curve and winds up losing the race.

Go-Moku

Go-Moku (variously translated as "five stones," "five squares," or "five 'eyes'") is a Japanese game for two players. A version called "Take 5" (using plastic pegs) has recently appeared in American toy stores. The original version of the game uses the same equipment as is used in the Japanese game of Go—a square-ruled board, usually 18 squares in length and width, and a supply of black and white counters (marker stones). In both Go and Go-Moku, players take turns placing counters (one at a time) on intersections at which lines of the ruling cross each other—that is, not inside the cells but at their corners (the longest horizontal row or vertical column can thus contain 19 counters). One player uses the black counters and the other the white ones, and once a counter is placed it cannot be moved again. In order to win in Go-Moku a player must place 5 counters in an unbroken line—vertically, horizontally, or diagonally.

The game is easily adapted to square-ruled paper. Instead of placing black and white counters on a board, the players draw on the board full (•) and open (○) circles, or better still, circles and crosses as in tic-tac-toe. It is also more convenient to put the marks in the centers of cells instead of putting them in corners: this does not change anything essential about the game, except that the Go "board" now ought to measure 19 by 19 cells. However, players often omit boundaries altogether and play with no limits on the available area. In any case, they take turns in making the opening move.

As was noted, the winner is the player who first establishes a continuous line of five marks, containing no empty spaces and no marks belonging to the opponent. Obviously, when a player has built up a line of, say, three marks, the opponent may place a mark at one end of this line to prevent it from growing at that end. A player who has established a line of four marks with "open" ends—no enemy mark at either end—is usually assured of victory: if the opponent cannot win in the next move, the best he can do is block *one* end of the line-of-four (see the illustration), which does not prevent the first player from winning by placing a mark at the *other* end.

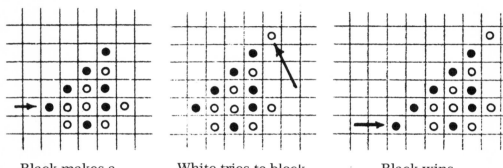

Black makes a winning move.   White tries to block.   Black wins.

The placing of a single mark that completes two lines-of-three, each with open ends, also brings victory in most cases (see drawing). If one developing line is blocked by the opponent, the other line can still be extended to a length of four and—unless the line is now blocked or the opponent can immediately win—there is no way of preventing its extension to five at the following turn.

White assures victory by completing two lines-of-three.

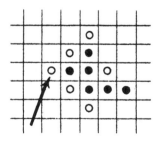

The related game of Go can also be played by marking the intersections (or the cells) on square-ruled paper: the object of the game, however, is not the establishment of a line-of-five but the surrounding of areas occupied by counters (or marks) of the opponent. It is a much more difficult game than Go-Moku and it contains many long-range traps and tactics, requiring careful planning of the game many moves ahead. In Japan, Go occupies a place similar to that of chess in Western society and there are nine recognized levels of expert Go players. Because of the complexity of Go, it will not be discussed here (you might find more about it from books devoted to games). Go-Moku, on the other hand, is a popular game, widely played by children and adults—often on square-ruled paper, in a manner similar to the one described here.

21

Random Walks

A "random motion" is one in which the direction keeps changing unpredictably. A kitten lost in a large open field will wander here and there, in directions which seem to be determined mostly by pure chance. A fish in the ocean, likewise, will often move "randomly"—without any preferred direction, as if every new move was decided by a "toss of the dice."

We can easily imitate such a "random walk" with the help of square-ruled paper, a pencil, and a coin. To make things simple, let us first look at a random walk *along a line*. The line will be the x axis, suitably marked

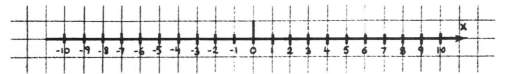

and we start out—naturally—from the origin ($x = 0$).

We now toss the coin: if it falls "heads" then x is increased by 1, if "tails" it is decreased by 1. The numbers one gets for x as the coin is thrown again and again give the position of a point randomly "walking" along the x axis: if you wish, you can note the changing position of this point by a pencil mark or a snippet of paper. Here are some "random walks" obtained in this way, each containing 16 "steps":

$$0, -1, 0, 1, 2, 1, 2, 3, 2, 1, 0, -1, 0, -1, -2, -1, -2$$

$$0, 1, 0, 1, 2, 3, 2, 3, 2, 1, 2, 3, 2, 3, 4, 5, 6$$

$$0, 1, 2, 1, 0, 1, 2, 1, 0, -1, 0, 1, 0, -1, 0, 1, 0$$

As can be seen, the progress of the point along the axis is completely unpredictable, and it is equally likely that x ends up positive or negative. In the last "walk" the point even returns to the origin, although this does not seem to happen very often.

After a certain number of steps, *how far from the origin*, on the average, does such a "random walk" extend? It will be seen that the result depends (partly) on the way in which we choose to define and calculate this "average"; and to make matters simple we will concentrate on random walks which have exactly 4 steps each.

At each step—each toss of the coin—there exist two choices for the change of x: either $+1$ or -1. This allows us to represent each completed 4-step random walk by a row of plus signs and/or minus signs. For instance,

would represent a walk in which the first 3 steps were equal to $+1$ while the last one was -1. Using this notation, one finds that there exist exactly 16 different random walks of 4 steps each, all of which are listed in the table that follows. For convenience, all walks which give the same total change in x are grouped together. This change is denoted by X and can be positive, negative, or zero (if we imagine the walk starting from $x = 0$, then X is simply the final value of x).

In addition, the table lists the "total distance covered": the distance between the starting point and the end point of the walk. In everyday conversation, when we speak about "distance" we always mean something positive (or just possibly, equal to zero), and the distances given in the table indeed follow this rule.

Table of Four-Step Random Walks on a Line

X, the total change in x	The steps of the random walk	$\|X\|$, the total distance covered	X^2
$+4$	$++++$	4	16
$+2$	$-+++$ $+-++$ $++-+$ $+++-$	2	4
0	$--++$ $-+-+$ $-++-$ $++--$ $+-+-$ $+--+$	0	0
-2	$+---$ $-+--$ $--+-$ $---+$	2	4
-4	$----$	4	16

You will see, if you check, that among the 16 walks listed, 8 have $+1$ as their first step and 8 have -1, and that the same equality also holds for any other step. It is also clear that X is equally likely to turn out positive or negative.

All 16 choices are equally likely to happen. If we carried out a very large number of 4-step walks, the irregularities caused by chance would tend to smooth out, and any particular combination such as $(+++-)$ would occur in very nearly 1/16 of the walks, the same proportion we get in the table by listing every case *exactly once*. So let's look at this set of 16 walks—in which each kind appears just once—and see how they behave "on the average"; this should be similar to the way a very large number of walks (selected not from a table but by pure chance) would behave.

The "ordinary average" or "mean" of X—denoted by pointed brackets, thus: $\langle X \rangle$—is the sum of the values of X from all walks, divided by the total number of walks being examined, namely 16:

$$\langle X \rangle = \frac{1}{16}\left[(4 \times 1) + (2 \times 4) + (0 \times 6) + (-2 \times 4) + (-4 \times 1)\right] = 0.$$

The result is zero, as expected, since X is equally distributed on the positive side and the negative side of the origin ($x = 0$).

We might also want to know, however, *how far from the origin* the average walk extends, and one way of achieving this is by using *absolute values*. The absolute value $|X|$ of a number X is just the positive number one gets after lopping off the sign. For instance,

$$|2| = 2, \text{ and}$$
$$|-2| = 2.$$

In our table, $|X|$ represents the total distance covered by the random walk, regardless of direction. Since $|X|$ is either positive or zero, its average will be a positive number. Can you calculate it? The calculation is very similar to that of $\langle X \rangle$, and you may compare your result to the answer on page 81.

There exists still another way of estimating the total distance covered by the random walk—by calculating the average of X^2, which like $|X|$ is always positive:

$$2^2 = 2 \times 2 = 4, \text{ and}$$
$$(-2)^2 = (-2) \times (-2) = 4. \qquad \text{("Minus times minus gives plus.")}$$

Averaging over the 16 cases listed in the table, we find that for our 4-step random walk the "average square of the distance covered" is

$$\langle X^2 \rangle = \frac{1}{16}\left[(16 \times 1) + (4 \times 4) + (0 \times 6) + (4 \times 4) + (16 \times 1)\right] = \frac{64}{16} = 4.$$

The square root of this should measure some sort of "average distance covered," and it amounts to

$$\sqrt{\langle X^2 \rangle} = \sqrt{4} = 2.$$

This is often written

$$\langle X \rangle_{\mathrm{rms}} = 2,$$

where "rms" stands for "root mean square," the mathematical name for this type of average. It does not equal the average $\langle |X| \rangle$ described earlier, which confirms that different ways of defining and calculating "averages" do indeed lead to different results.

Which of these two averages is more useful? It turns out that the r.m.s. average is more easily generalized for any number of steps, of any size (provided all steps are still of equal size). If the random walk contains N such steps, we find

$$\langle X^2 \rangle = N \times (\text{size of step})^2.$$

(Could you check this for $N = 1$?) And from this we get

$$\langle X \rangle_{\mathrm{rms}} = \sqrt{N} \times (\text{size of step}).$$

For example, if each step equals either $+3$ or -3 and 100 steps are taken, then according to the formula the random walk will on the average (the r.m.s. average!) take us 30 units away from where we started.

A random walk which covers an *area* (like the walk of the lost kitten) is easily drawn on square-ruled paper with the help of a set of (x,y) coordinate axes. As before, you start out at the origin, but now at each step you throw *two* coins—say, a dime to give the change in x and a penny to give the change in y. One such random walk, containing 49 steps of equal length, is shown here: because only 4 directions are available, the walk will often retrace itself and may even go around closed loops.

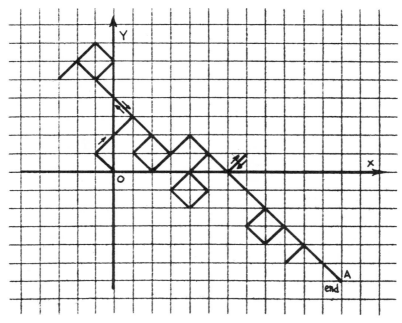

Each step in this walk advances us *diagonally across* one of the squares of the ruled paper, and its length is therefore always $\sqrt{2}$ (see chapter 7). However, the formula for the r.m.s. average distance between the starting point and the end point, after N steps, still remains

$$\langle X \rangle_{\text{rms}} = \sqrt{N} \times (\text{size of step}) = \sqrt{N} \times \sqrt{2}.$$

This distance will not be, in general, along one of the main axes—in the diagram of the 49-step walk, if the origin 0 is viewed as the center of a clock, the direction from it to the end point A points roughly toward 4 o'clock. You may check for yourself how close the distance OA comes to the average value derived from the formula, using the edge of a sheet of square-ruled paper as a ruler and remembering that each step equals $\sqrt{2}$ or approximately 1½ units of the ruling.

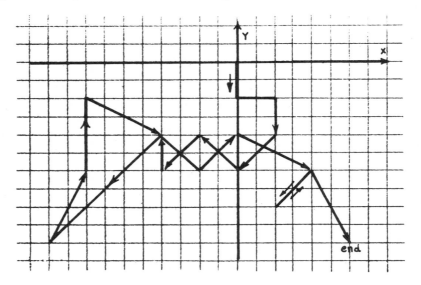

You can trace a random walk faster if you throw several pairs of coins at the same time: in the drawing of the second random walk shown here, 4 dimes (for changes in x) and 4 pennies (for changes in y) were thrown together at each step. Each step therefore represents the total effect of 4 "ordinary" steps—which means, if you think it over, that each change of x or of y is now the result of one of the 4-step random walks listed in our table and can equal 4, 2, 0, -2 or -4. Therefore there are more possible directions for the walk, and the steps also may differ in size, making it less likely for the walk to retrace itself (although it *does* happen here, just before the end).

There are 25 steps altogether, including 5 "invisible" ones in which the change in both x and y turns out to be zero. As noted, each step is equal to four "ordinary" steps, like those of the 49-step walk: instead of tossing 4 pennies and 4 dimes all at once, we could start by tossing the first dime and the first penny to get the first "ordinary" step, then repeat with a second pair of coins, then with a third and a fourth. Thus the 25 "large" steps represent 100 "ordinary" ones, and according to the formula the average r.m.s. distance covered in 100 such steps is

$$\langle X \rangle_{\mathrm{rms}} = \sqrt{100} \times \sqrt{2} = 10 \times \sqrt{2},$$

or about 14. Measure for yourself how close this predicted average comes to the actual distance covered by the random walk!

Finally we take a quick look at random walks in three dimensions. Air and other gases consist of tiny bits of matter, called *molecules*, which fly through space and collide with each other many thousand times each second. Each time a molecule collides, the direction in which it moves changes unpredictably, so that it moves in a kind of three-dimensional random walk. The steps in this "walk" do not have equal sizes, because after any collision one can never tell how far a molecule will move before it collides again.

When perfume is left in an open bottle, some of its molecules mix with the surrounding air and perform similar "random walks": each collision with an air molecule launches the perfume molecule into a new and unpredictable direction. After many bounces, a few such molecules may even reach our noses, where their presence is noted. One says that molecules of perfume *diffuse* through the air (in addition, air currents help in spreading them around).

Our factories are the source of many kinds of molecules which diffuse through the Earth's atmosphere and which affect the environment in which we live. Of special interest are molecules belonging to a family of gases known as "Freons," which are used in refrigerators, air conditioners, and certain kinds of spray cans. Freon molecules do not break up easily: one of the few things that can tear them apart is ultraviolet light, a "color" of light not visible to the eye (though it *does* affect photographic film). About 1/100 of the Sun's light output is of this kind, but luckily for us most of it is blocked by a layer containing a small amount of ozone—a special form of the oxygen we breathe—which is produced high in the atmosphere and is mostly found at heights between 6 and 20 miles. Very little of the Sun's ultraviolet light manages to get

Solution

The average of $|X|$:

$$\langle |X| \rangle = \frac{1}{16} \left[(4 \times 1) + (2 \times 4) + (0 \times 6) + (2 \times 4) + (4 \times 1) \right] = \frac{24}{16} = 1\frac{1}{2}.$$

through the ozone layer and to reach us; and that's a good thing, because ultraviolet light strongly tans and even burns the skin, and is in addition unhealthy to the eyes.

The trouble is that the ozone layer acts as a shield not only for us but also for the millions of tons of Freon which escape every year from broken air conditioners or from spray cans. The released Freon molecules diffuse randomly through the atmosphere and nothing much happens to them unless they happen to reach above the ozone layer, into the region where the Sun's ultraviolet light can break them apart. Unfortunately, when this happens, one of the fragments is the gas chlorine which destroys ozone quite efficiently. If too much ozone gets destroyed in this way, an unhealthy amount of ultraviolet light will be able to reach the ground.

It takes on the average about 20 years of random-walking before a Freon molecule wanders to where sunlight can destroy it, so gases which are released today may cause their damage only far in the future. Still, many people have become concerned about the possibility of such damage, and the U.S. government has accordingly banned the use of Freon in most kinds of spray cans.

22

Families of Points

Interesting things are seen when one collects "families" of points for which the values of the x coordinate and the y coordinate are in some way related. Take for instance the points for which x equals y. These include

$$(1,1), (2,2), (3,3), (4,4), (5,5),$$

and, of course, many others. If you draw these points you will note at once that they all fall on a straight line.

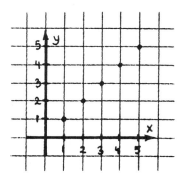

This straight line also passes the origin (0,0)—and there, too, x equals y, both of them being zero. It also passes such points as

$$(-1,-1), (-2,-2), (-3,-3)$$

and even

$$(\tfrac{1}{2},\tfrac{1}{2}), (2\tfrac{1}{2},2\tfrac{1}{2}), (-3\tfrac{1}{2},-3\tfrac{1}{2}).$$

All of these have been added to the drawing shown here:

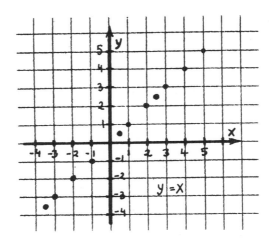

In short, one finds that the line passes *all* points for which the number x equals y, regardless of whether this number is positive, negative, or zero, or whether it is whole or fractional. Naturally, we call it

"the line of $y = x$."

Similarly, "the line of $y = 2 \cdot x$" includes all points for which y is *twice* as big as x. In order not to confuse x with the multiplication mark we will from now on denote multiplication *with a dot*, as mathematicians often do. To avoid confusion with the decimal point, a little extra space is left in front of the multiplication dot and behind it, and it is sometimes lifted slightly above the line. Computer programmers go even further towards making the multiplication symbol distinctive and unique by using the asterisk—that is, "two times three" is written 2 * 3—a practice which perhaps deserves to be adopted by schools.

Coming back to our families of points: the line of

$$y = 2 \cdot x$$

includes points with y equal to two times x, such as

$$(1,2), (2,4), (3,6), \ldots, (-1,-2), (-2,-4), \ldots.$$

Again, the origin $(0,0)$ also belongs to the family since for it, too, y is equal to twice x. Because the line happens to be straight, we only have to find two points on it: once these are marked we can use a ruler to draw the line itself. Here is how the line will appear:

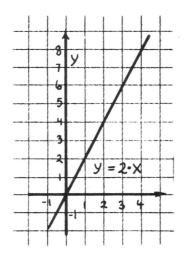

The line $y = 3 \cdot x$ is drawn in much the same way—using, for instance, the points (1,3) and (−1,−3) or (3,9) as guide marks for the ruler.

Note that such a line can be used for multiplying numbers by 3, since the value of y at any point is always 3 times the x coordinate. For instance, to get $4 \cdot 3$ (the dot marks multiplication, remember!) we look for the point on the line which has $x = 4$ (see drawing). The value of y for this point is 12, and this indeed equals 4 times 3.

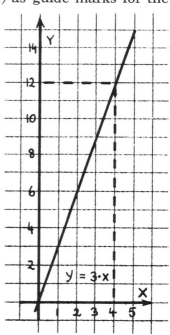

Another straight line includes points with x and y related by

$$y = \tfrac{1}{2} \cdot x$$

and it has the following appearance:

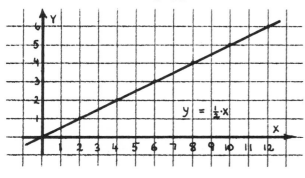

It includes points such as

$$(0,0), (1,\tfrac{1}{2}), (2,1), (3,1\tfrac{1}{2}), (4,2), \ldots, (-2,-1), (-1,-\tfrac{1}{2}).$$

The line

$$y = \tfrac{1}{3} \cdot x$$

has this form:

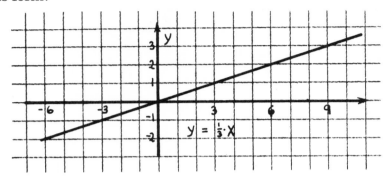

It passes the points $(0,0)$, $(3,1)$, $(6,2)$, . . . , $(-3,-1)$, $(-6,-2)$, and an infinite number of others.

There are of course many other lines (infinitely many, indeed) which can be drawn in this way: you may experiment, if you wish, with these:

$$y = 1\tfrac{1}{2} \cdot x.$$
$$y = -x.$$
$$y = 3 + x.$$
$$y = 3 - x.$$

All these are straight and may therefore be drawn with a ruler as soon as *two* points on them have been marked. *Be warned*, however, not to place these points too close to each other—otherwise, if the line misses one of the marks by even a small amount, the resulting error is quite noticeable!

23

Parabolas

Not all families of points give straight lines. Take for instance the points with

$$y = x^2.$$

It helps things if, before marking points on paper, one first draws up a *list* of values of x and y that belong to each other. Let x be chosen from the lowest whole numbers: 0, 1, 2, 3, and so forth. The values of y are then the squares of these numbers, so we get:

If x is	0	1	2	3	4	5
then y is	0	1	4	9	16	25

If you are familiar with the rule "minus times minus equals plus," you can at once add to the table points with negative x. For instance, if $x = -3$, then (remember, dot means multiplication)

$$y = (-3)^2$$
$$= (-3) \cdot (-3) = +9.$$

One therefore gets:

If x is	-1	-2	-3	-4	-5
then y is	1	4	9	16	25

Drawing all these points and connecting them smoothly gives a curve called a *parabola*.

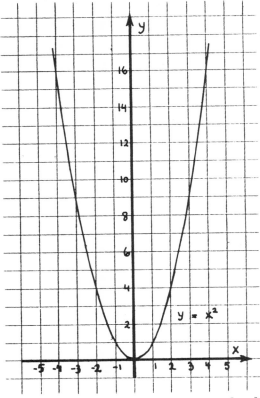

If drawn correctly, the parabola will pass not only through the listed points but also through all in-between points for which $y = x^2$. For instance,

$$(x,y) = (\tfrac{1}{2}, \tfrac{1}{4})$$

is such a point.

Parabolas can be found in many places. In a suspension bridge, where the roadway hangs from towers by means of long cables, these cables usually form the bottom of a parabola. (Please note, however, that if the cable were to be left hanging alone, without carrying a roadway, its shape would be slightly different.)

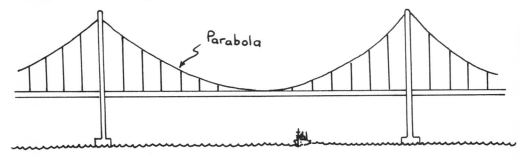

A stone thrown upward at an angle moves along an upside-down parabola. So does a cannon shell or a baseball—more or less: the curve may differ slightly from an exact parabola, because the air resistance that slows down the motion also slightly changes the shape of the curve.

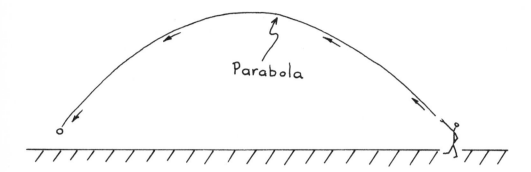

The shapes of the orbits of many comets that arrive near the sun from deep space are also very similar to parabolas. Since a parabola is open at its ends, we can assume that such comets make only one pass at the sun and never come back.

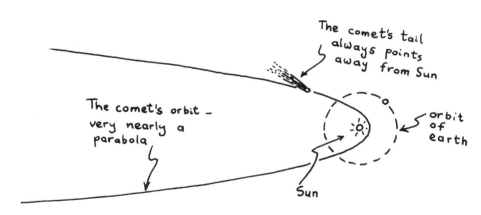

The curves representing relations such as

$$y = 2 \cdot x^2$$
$$y = \tfrac{1}{2} \cdot x^2$$

are also parabolas, but of narrower or wider shape than the one we drew earlier.

How about deriving a family of points from *triangular numbers?* Suppose x and y are related through

$$y = \Delta_x.$$

That is, y is to be the x-th triangular number (x obviously must be a whole number). As before we begin by drawing up a list:

If x is	1	2	3	4	5	6	7
then y is	1	3	6	10	15	21	28

If you mark the points in (x,y) coordinates and connect them with a smooth curve you will get just about half a parabola. In fact, the points all belong to the curve

$$y = \tfrac{1}{2} \cdot x \cdot (x+1),$$

which *is* a parabola. This should not come as a surprise, since the formula for triangular numbers is

$$\Delta_N = \tfrac{1}{2} \cdot N \cdot (N+1).$$

Therefore, in the relation which defines the curve (and we are going to discuss many other such relations in the next chapter), whenever x equals some whole number N, then the curve's equation is the same as

$$y = \Delta_x.$$

When x is not whole, Δ_x has no meaning, but one can still find a point on the curve $y = \tfrac{1}{2} \cdot x^2 \cdot (x+1)$. If you draw that curve you will find that it is very similar to the parabola

$$y = \tfrac{1}{2} \cdot x^2$$

but with a shift downwards and to the left: the parabola $y = \tfrac{1}{2} \cdot x^2$ (like the one first drawn) "hits bottom" at the origin, but the other curve bottoms out at $(-\tfrac{1}{2}, -\tfrac{1}{4})$. By drawing $y = \tfrac{1}{2} \cdot x \cdot (x+1)$ on one sheet of paper and $y = \tfrac{1}{2} \cdot x^2$ on the other, holding both sheets together against a bright light, and then sliding one sheet over the other until the curves overlap, you can convince yourself that they do indeed have the same shape.

24

Graphs

So far, coordinates have been used for producing two kinds of lines. The simplest relations, such as

$$y = x$$
$$y = 2 \cdot x$$

gave *straight* lines, while relations involving squares, like

$$y = x^2$$
$$y = 2 \cdot x^2$$

gave *parabolas*. More generally, it may be shown that if a, b, and c are any given numbers, all relations of the form

$$y = (a \cdot x) + b$$

give straight lines (for this reason these are called *linear relationships* between x and y), while those of the form

$$y = (a \cdot x^2) + (b \cdot x) + c$$

give parabolas. However, there exists no limit to the number of different curves that can be drawn by this method. One can invent many different rules connecting x and y, and each rule (or "equation") has its own family of points. Look at these, for instance:

$$y = \frac{1}{x}.$$
$$y = x^3 - (5 \cdot x) + 3.$$
$$y = \sqrt{x}.$$

The points described by such rules can generally be connected by smooth curves of various shapes. Some of the curves wiggle and snake up and down. One of the rules listed here, though, gives a parabola: can you guess which one? If you turn to page 94, you can check your guess against the solution.

Lines of the type discussed here are called *graphs*: a large part of mathematics has to do with graphs, and scientists in all fields use them frequently. Square-ruled graph paper which is finely divided into small squares—such as the paper with squares one millimeter apart, shown in chapter 16—is particularly useful for this purpose. Nowadays it is fashionable to talk casually about millions or even billions of people, dollars, and so forth. Graph paper with squares one millimeter wide can make us realize how big a million actually is: over 20 page-size sheets of it are required for presenting a total of one million squares. To display one million squares with the graph paper used in most of this book, about 500 pages would be needed!

By the way, you may have noted that in all the examples given so far, the complicated part of the rule involved x. It is the custom in mathematics to leave y alone as much as possible. Sometimes this rule must be broken, and at the end of this chapter an example will be given in which it is more convenient *not* to set y apart.

Just as there exists a special name—"graph"—for the line that connects points of a given family, so there also exists a special name for the *rule* relating y and x in any such family of points. If such a rule exists, mathematicians say that "y is a *function* of x" (one could equally well say that x is a function of y, but we have already noted that by tradition y is the quantity usually isolated). For example, in any of the relationships shown earlier in this section, y is *some* function of x.

There exist *two kinds* of graphs (and of functions).

The kind we have been dealing with thus far are graphs in which the relation between x and y is purely mathematical. Examples:

$$y = x.$$
$$y = x^2 + x + 1.$$
$$y = 7 - (2 \cdot x).$$
$$y = \frac{12}{x}.$$

The other kind includes graphs that describe relations that are not calculated but *measured*. For instance, x could be the time in years as counted on the calendar, while y would be the number of people living at that time in the United States. Let us take a closer look at this particular example.

By the law of the Constitution all people in the U.S. are counted every 10 years (this is called a *census*). Listed below are the results of censuses since 1810, with y the population of the United States in millions, rounded off to the nearest million:

x	1810	1820	1830	1840	1850	1860	1870	1880	1890
y	7	10	13	17	23	31	39	50	63

x	1900	1910	1920	1930	1940	1950	1960	1970
y	76	92	106	123	132	151	179	205

It is quite easy to draw a graph of this relation between x and y if we take care to adjust it to fit the page (one price paid for this adjustment is that the origin is no longer included in the graph). We choose every square in the x direction to represent 5 years while every square in the y direction stands for 10 million people. Drawing the graph itself might require a certain amount of "reading between the lines": to mark 7 millions, for instance, we must guess or measure a distance which is 7/10 of the width of a square.

In this manner we get the following graph:

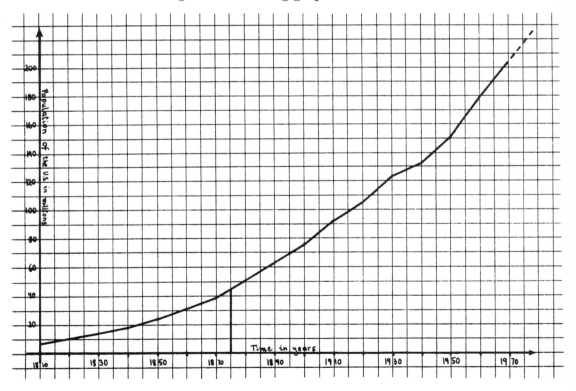

One immediately sees that the growth started slowly and became faster as time went on. This is easily understood—more children are born in a nation of some 200 million people—as we were around 1970—than in a nation of one-fifth this size, as we were a hundred years earlier. Yet this is not the complete story—the U.S. also grew in size during those years, as new states joined the union, and many of the added people were not born here but arrived from other countries. In the 1930s, conditions of life in the United States became relatively hard ("The Great Depression"). This caused a decrease in the birth rate—the average number of children born each year, for every 1000 people—and at the same time the number of new arrivals dropped sharply, as a result of new immigration laws and the difficulty of finding work in the U.S. Indeed, the graph shows that the rate of growth was noticeably smaller during those years.

Population is counted every 10 years, but with a graph one can also guess rather accurately how many people there were in the in-between years. To find the U.S. population in 1875 (for instance) we draw a vertical line at $x = 1875$ and mark the point at which it hits the curve (see drawing!). The value of y at this point then approximately gives the population in 1875. Such "reading between the points" is called *interpolation*.

More interesting, we can also try and continue the graph into the future (broken line in the drawing) and thus make a guess at how large the population will be in 10 or 20 years. Such "reading past the end of the

---

Solution

The rule for a parabola:

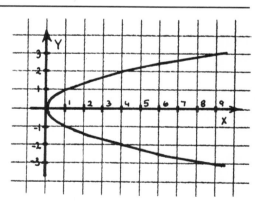

Among the rules listed, the third one implies that $x = y^2$. It therefore represents the same relationship as the one which gave the parabola drawn in the preceding section, except that the roles of x and y are now exchanged. Because of this exchange, the parabola is "lying on its side"—but otherwise its shape is exactly the same.

Note that, with this rule, for every x there exist *two* possible choices of y. For $x = 4$, for instance, y equals $\sqrt{4}$ or "the number the square of which equals 4," which can be *either* 2 *or* −2. Cases like this one, in which more than one correct answer exists, are often met in mathematics.

graph" (*extrapolation*) assumes that things will continue in the future the way they have been in the past: for predicting the future a short way ahead it works well, but trusting it too far past the last measured point can quickly bring large errors.

Mark Twain, in *Life on the Mississippi*, poked fun at "scientific people" who extended their observations too far past their last point, or ahead of the first one; and he proposed some rather wild extrapolations of the length of the Mississippi river. That river, along much of its length, winds left and right in big curves, making it appear (as Twain put it) as "the crookedest river in the world." Such windings of the river (along the eastern edge of Arkansas, for instance) may make it twice as long as it would have been if it followed a straight line.

Now and then, after a flood, the river is found to have shortened itself by taking a shortcut across one of its curves. Twain claimed that over the 176 years preceding his own time the river had shortened itself, on the average, by more than a mile per year. He then extrapolated this trend to ridiculous extremes—to the far future, when (if the same rate continued) the river's length would shrink to nearly nothing, and to the distant past, ahead of the beginning of the graph, when the river's length, assumed to increase by about one mile with every additional year one went back in time, was far to big to fit the surface of the earth:

> Now, if I wanted to be one of those scientific people, and "let on" to prove what had occurred in the remote past by what had occurred in a given time in the recent past, or what will occur in the far future by what has occurred in late years, what an opportunity is there! . . . Please observe:
>
> In the space of one hundred and seventy-six years the Lower Mississippi has shortened itself two-hundred and forty-two miles. That is an average of a trifle over a mile and a third per year. Therefore, any calm person, who is not blind or idiotic, can see that in the old Oolitic Silurian Period, just a million years ago next November, the Lower Mississippi River was upward of one million three hundred thousand miles long, and stuck out over the Gulf of Mexico like a fishing rod. And by the same token any person can see that seven hundred and forty-two years from now the Lower Mississippi will be only a mile and three-quarters long, and Cairo and New Orleans will have joined their streets together, and will be plodding comfortably along under a single mayor. . . . There is something fascinating about science. One gets such wholesale returns of conjecture out of such a trifling investment in fact.

Back now to the graph, which could be called "a graph of U.S. population against time." Other measured relationships may also be expressed in graphs—the graph of the distance y needed to stop a car, against the speed x at which the car travels (if you used the *formula* given in chapter 5 for this relationship, by the way, you would end up with a parabola), or of the average height or weight of a person against his or her age, and so on.

In all cases a graph provides at a glance many details about the relationship it describes. A look at the parabola $y = x^2$, for instance, immediately shows which x has the smallest y connected with it—this happens at the origin, where the curve hits its lowest point.

Scientists working with computers often make them give the results of calculations by means of graphs, for they can then see easily and quickly what these results mean. Usually a computer prints out its results by means of an electric typewriter which can produce not only level dashes (like ——) but also vertical ones (like ||||). This enables the machine to print out both axes, as well as the numbers labeling them and the points of the graph, which are marked by an × or some similar symbol, as in the illustration.

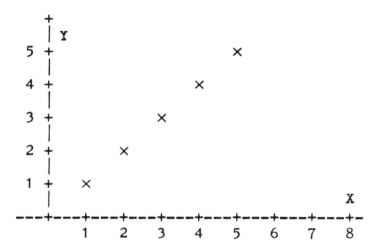

This is not completely satisfactory, because a typewriter can move sideways only a whole number of spaces and up or down only a whole number of line-spacings. To overcome this there exist special writing machines, containing a pen which moves above a sheet of paper: by means of pulleys and motors the computer can move the pen all over the sheet and make it draw quite complicated designs and even write down the numbers and letters to label the graph. A different method has the graph "drawn" on the screen of a picture tube somewhat similar to the one used in television receivers, and there are also ways of copying it on a sheet of special paper.

The word *graph*, by the way, comes from a Greek word meaning "to write." Indeed, a graph is a way of "writing down" the relationship between changing quantities.

This word has several relatives in our language. *Graphology* is the study of handwriting—experts in graphology claim that handwriting tells a great deal about a person's true character. *Graphite* is a form of carbon (the black substance in coal and charcoal) which is used in the writing part of a pencil—a part still called "the pencil's lead" because long ago the soft metal lead was used for it instead of graphite.

In 1844 the American inventor and painter Samuel Morse introduced a machine that could write down signals sent to it electrically. He named it *telegraph*—the machine that writes (a message) at a distance (*tele* is "far" in Greek). Just as *graph* means writing, *gram* means that which is written, so *telegram* is the message sent by telegraph, and *grammar* is the study of rules to be followed when writing.

To conclude this chapter, here is a problem:

Can You Draw This Graph?

Suppose that x and y are related by the condition

$$y^2 + x^2 = 25.$$

In this case it is more convenient to leave the condition in its original form rather than try to separate y to "stand by itself." What is then the shape of the graph passing through all points at which the above relation between x and y holds true?

To help you out, here is a table of some points belonging to the graph:

If x is	0	3	3	4	4	5	0	−3	−3	−4	−4	−5
then y is	5	4	−4.	3	−3	0	−5	4	−4	3	−3	0

Note that for every x there may exist *two* points on the graph (just as we found in the "sideways parabola" of the preceding problem): for instance, for $x = 3$ we may have either $y = 4$ or $y = -4$. Because of the multiplication rule "minus times minus equals plus," both values have the same square:

$$4^2 \ = \ 4 \cdot 4 \ = 16.$$
$$(-4)^2 = (-4) \cdot (-4) = 16.$$

Therefore in both cases the condition of the graph is fulfilled: that is,

$$3^2 + y^2 = 25.$$

Try to draw the graph, and then compare your result with the answer on page 100—but don't look before you have tried!

25

The Slope of a Line

Let us examine two "roof-shaped" triangles, each of them three units high:

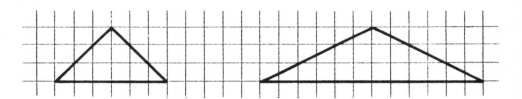

The triangle on the left is *steep*: it would be hard to climb a roof of such a shape without slipping. The one on the right is *less steep* and has a much more "climbable" shape.

How does one measure steepness?

As one climbs the side of the steep triangle, with every unit that one advances to the left or to the right one also rises one unit. Going from the bottom to the top one rises 3 units and at the same time one also advances 3 units horizontally.

As one climbs the side of the less steep triangle, one only rises *half* a unit for every unit advanced horizontally.

It is customary to say that the first triangle has a *slope* of 1, the second a slope of ½. More generally, the slope of a slanting straight line is the number of units by which the line rises for every unit advanced horizontally. The word "unit" here may mean *any* unit of length, be it the length of a square on our ruled sheet of paper or be it an inch, foot, or meter: no matter which units are used, the slope of a given line is always the same.

The slope provides an accurate way of measuring steepness: the larger it is, the steeper is the line. Some examples are drawn below:

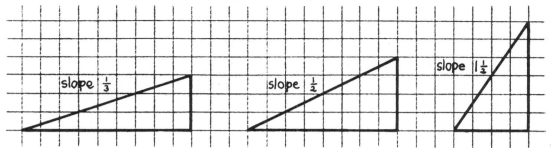

There is a simple rule for calculating slopes: if we rise *A* units while advancing *B* units horizontally, the slope equals the fraction *A/B*. For instance, the side of a triangle which rises 2 units while advancing 5 units has a slope equal to 2/5 (two fifths).

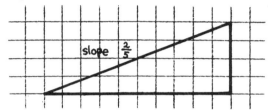

Highway builders are obviously interested in the steepness of roads. What we here called the slope they call the *gradient* or *grade* of the road, and it is usually measured in *percent*, that is, in hundredths. A slope of 8 percent—written 8%—means that for every 100 feet that the road advances, it rises 8 feet (those 100 feet are sometimes measured along the sloping road itself instead of horizontally, but for small slopes this makes very little difference). Such a slope may seem rather mild if drawn on ruled paper, but actually it is a fairly steep one for a road.

Driving a car on a steep road demands extra care—especially when you are driving downhill, for the slope then makes it difficult to stop the car quickly. In the United States, steep stretches of road usually begin with yellow warning signs bearing messages such as CAUTION—STEEP ROAD AHEAD or often simply HILL.

In most of the world there is a different system of traffic signs, using *pictures* instead of words—perhaps because often not all the drivers on a road speak the same language. Warning signs, in particular, are triangle-shaped with red borders and contain a picture which explains their purpose. For instance, there are signs warning of sharp curves ahead, or of railroad crossings.

The warning sign for a steep road shows a rising slope, and sometimes also tells what the grade of the road is.

A straight-line graph also has a slope: for instance, the straight graphs drawn below have slopes of ½, 1, 2, and 3. There is a simple rule connecting the slope of such a graph with the equation that describes it, but we'll leave it to you to guess it.

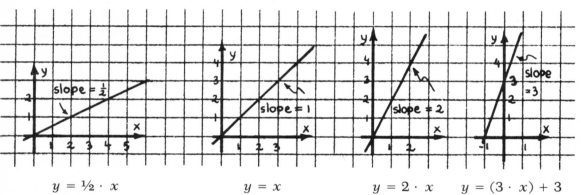

$$y = \tfrac{1}{2} \cdot x \qquad\qquad y = x \qquad\qquad y = 2 \cdot x \qquad y = (3 \cdot x) + 3$$

Solution

"Can You Draw This Graph?" (page 97):

The graph is a circle, 10 units wide and centered on the origin.

26

The Puzzle of the Extra Cell

Draw a square of 8 by 8 units and divide it into four parts like this:

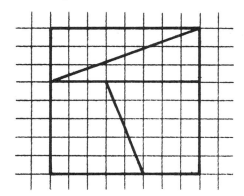

Here the parts are rearranged to make a rectangle 13 units long and 5 units high:

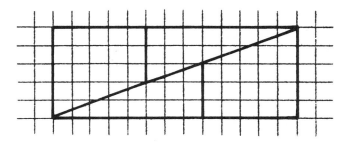

But is this really possible? The rectangle has an area of $13 \cdot 5 = 65$ cells, while the area of the square is only $8 \cdot 8 = 64$. If the same pieces were used for both, where did the extra cell come from?

Before you start your search, here is a hint: this puzzle has to do with slopes. You will need the formula for slopes that are given by fractions, examples of which were given earlier. For instance, if a line rises 3 units while advancing 8, its slope is given by the fraction 3/8.

(The full solution begins on page 104.)

27

The Steepness of a Curved Line

As the last stop on our mathematical trip, we will briefly discuss the steepness of a *curved* line. Unlike the steepness of a straight line, it usually changes from one point to another. Suppose this is the picture of a hill:

Near the bottom the hill is not very steep; but as one climbs, the ground grows steeper and steeper, until at the point marked A it has considerable slope. Then as one climbs still farther the steepness eases off again, and at the very top the hill seems quite level and one senses no slope at all.

Now suppose that the above drawing is *not* a hill but a curved line drawn on a sheet of paper—which, after all, *is* closer to the truth. Then all that was said about the slope of the hill at various points also holds true for the slope of the curve. The steepness of other curves is handled in much the same way: if the curve is a graph drawn in (x,y) coordinates, then the x axis is usually regarded as "level ground" and all slopes are compared to its direction.

Solution

The Puzzle of the Extra Cell:

If you cannot see anything wrong, you have been tricked. The pieces of the 8-by-8 square, with a total area of 64, only *seem* to make up a rectangle of 13 by 5 units. Actually they do not fit together completely in the way drawn.

Take the line through the 3 points labeled *A*, *B*, and *C* below: if the pieces all fit exactly into the rectangle, it should be a straight line.

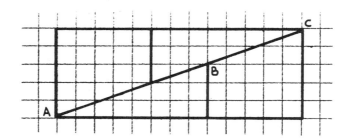

But is it straight? Let us examine the slopes of its two parts. A straight line from *A* to *B* (we will call it "the line *AB*" or just "*AB*," for short) rises 3 units while advancing 8, so its slope is 3/8. The line from *B* to *C* (the line "*BC*") rises 2 units while advancing 5, so its slope is 2/5. We may always multiply the top and bottom of a fraction by the same member (5 or 8 in this case) so:

$$\text{Slope of } AB = \frac{3}{8} = \frac{15}{40}.$$

$$\text{Slope of } BC = \frac{2}{5} = \frac{16}{40}.$$

The slope of *BC* is larger—which means that *BC* is *steeper* than *AB* and that the combined line really looks something like the way it is drawn below (we exaggerate the two slopes to make it more clearly visible):

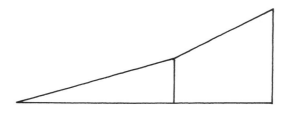

Solution (continued)

The two pieces along the top side of the rectangle also have different slopes, and the way everything *really* fits into the 13-by-5 area is shown below (again exaggerated):

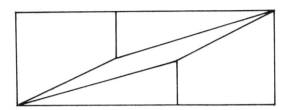

The pieces of the square do not completely cover the rectangle but leave a long, narrow slit in the middle, and that is where the extra area has come from. It is actually a very narrow slit, which explains why it was possible to "hide" it.

You can stump your friends with this puzzle by cutting out an 8-by-8 square and dividing it into four pieces which can be rearranged (or so it seems) into a 13-by-5 rectangle. If you are deliberately careless in cutting out the pieces so that their sides are not completely straight, no one is likely to notice that they really do not quite fit together.

To get a clearer idea of what is meant by "the slope of a curve near some point P on it," one draws through P a straight line which follows the direction of the curve near P:

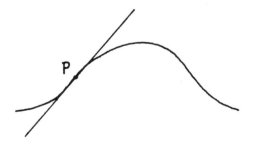

This line is called the *tangent* to the curve at the point P. Because the tangent has the same *direction* as the curve near P, it also has the same *slope* as the curve has there. While it is difficult to measure the slope of the curve at P, it is rather easy to measure the slope of the tangent, since it is a straight line. Therefore we can easily find the slope of the curve at P—*if* only we can draw a tangent there!

Unfortunately, this is not so simple. If you try drawing a tangent to a smooth curve you quickly find that it is rather hard to decide (even if you have a transparent ruler) which is the *exact direction* of the curve at some given point P on it.

As an aid in drawing tangents, engineers sometimes use a small rectangle-shaped mirror. First, place the drawing of the curve on a table and stand up the mirror vertically at the point at which the tangent is to be drawn—the point marked P in the picture shown here:

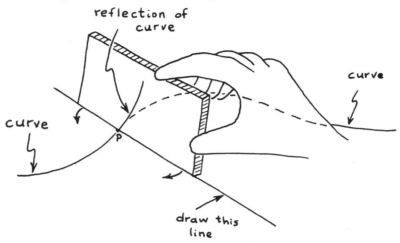

One must now experiment with the mirror, turning it slightly around P in the directions of the small curved arrows, until the reflection of the curve in the mirror appears to be a *smooth continuation* of the curve drawn on the paper, with no corners or angles. When this happens, draw a pencil line through P, using the bottom of the mirror as a ruler.

Next the mirror is put away. With a drafting triangle or a similar instrument having a right-angle corner, a line is now drawn through P in a direction perpendicular to the line marked earlier along the mirror. This line is the tangent: it is shown in the drawing below, and one can use a ruler to continue it on both sides of P.

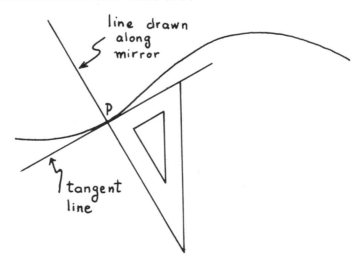

Of particular interest are tangents of graphs that describe a mathematical relationship between x and y. Take for example the parabola

$$y = \frac{1}{2} \cdot x^2.$$

We have drawn the graph of this relation three times to show the tangents to it at the points with $x = 1$, $x = 2$, and $x = 3$:

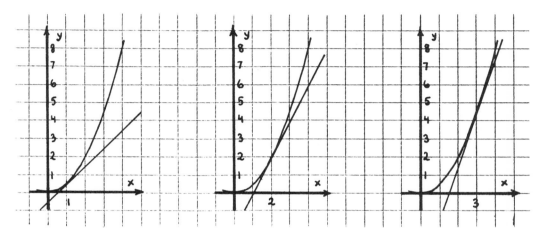

As the tangent lines show, the slopes of the graph at these points also equal 1, 2, and 3. It is easy to guess the simple rule which gives the slope of this particular parabola at any given point. You may note that this rule also applies to yet another tangent to the parabola—the x axis, which is the tangent at $x = 0$ and has zero slope.

As it happens, we are here but a short step away from a wide-ranging field in mathematics, called the differential and integral calculus ("calculus," to college students who study it), which deals with graphs and functions and the way the relations described by graphs grow and change.

Unfortunately we cannot take this step: we have skipped past too many mathematical tools which would be required for this. Perhaps some day, in a more systematic study of mathematics, you will be able to continue from where we now stop.

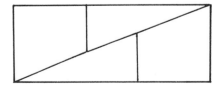

APPENDIX

Why $\sqrt{2}$ Is Irrational

Suppose $\sqrt{2}$ could be written as a fraction:

$$\sqrt{2} = \frac{a}{b} \ .$$

Let us add here the condition that a and b are the *smallest* whole numbers suitable for this purpose (a must be larger then b, since $\sqrt{2}$ is larger than 1). This condition is needed because any fraction can be written in many different ways, obtained by multiplying top and bottom by the same number; for instance ½ can be written

$$\frac{2}{4} \ \text{ or } \ \frac{3}{6} \ \text{ or } \ \frac{17}{34} \ \text{ or even} \frac{48651}{97302} \ \text{(using all ten digits),}$$

but the form ½ is the one using the smallest numbers.

If the fraction a/b satisfies this condition it cannot happen that a and b are *both* even, for if that were so, then smaller whole numbers m and n would exist such that

$$a = 2 \times m \text{ and } b = 2 \times n.$$

The fraction would then equal

$$\frac{a}{b} \ = \frac{(2 \times m)}{(2 \times n)} = \ \frac{m}{n} \ ,$$

and the form m/n would express it in numbers smaller than a and b.

However, a^2 *must* be even. To show why this is unavoidable we multiply the fraction by itself:

$$\sqrt{2} \times \sqrt{2} = \frac{a}{b} \times \frac{a}{b} \, .$$

Then, using the rule for multiplying fractions (top by top, bottom by bottom) gives

$$2 = \frac{a^2}{b^2} \, .$$

Multiplication of both sides by b^2 then shows that a^2 equals twice the whole number b^2 and is therefore even:

$$2 \times b^2 = a^2 .$$

Now, since a^2 is even, then a itself must also be even (if a were odd, a^2 results from multiplication of two odd numbers and is also odd). A whole number m must therefore exist such that

$$a = 2 \times m .$$

Then

$$a^2 = (2 \times m) \times (2 \times m)$$
$$= (2 \times 2) \times (m \times m)$$
$$= 4 \times m^2 .$$

We therefore get

$$2 \times b^2 = 4 \times m^2 .$$

Taking one half of each side of this equality gives

$$b^2 = 2 \times m^2 .$$

But this means that b^2 is also even, and by the arguments just used, b itself is then even too. This creates a problem: when we started we assumed that a and b were not both even, yet now we are forced to admit that they are.

This problem can only avoided if we realize that there is something basically wrong with the assumption that $\sqrt{2}$ can be written as a fraction. Suppose that a and b are indeed both even: then, as shown before, the fraction a/b can be represented in terms of smaller numbers m and n:

$$\sqrt{2} = \frac{m}{n}.$$

However, the same arguments which have been applied to a and b may now be used to show that m and n themselves are also even: the numbers m and n may then be replaced by numbers m' and n' half their size. This process can be repeated as many times as one pleases, each time both parts of the fractions being replaced by whole numbers half as large.

Clearly, no known fraction permits itself to be reduced infinitely many times. The only remaining possibility is then that *no* fraction exists which represents $\sqrt{2}$.

The proof given here was known to the ancient Greeks and may well have been the one which confounded the Pythagorean brotherhood, although this is not known for sure. With an accuracy of one millionth, the value of $\sqrt{2}$ can be written as the decimal fraction ($\cong$ means "approximately equal")

$$\sqrt{2} \cong 1.414214.$$

Index

Note: Pages listed in **bold face** type contain the definition of the indexed item. Pages listed in parentheses ( ) are those of problem solutions.